Developments in Modern Physics

DEVELOPMENTS IN MODERN PHYSICS

Nelson Bolívar

ARCLER PRESS

www.arclerpress.com

Developments in Modern Physics

Nelson Bolívar

Arcler Press
224 Shoreacres Road
Burlington, ON L7L 2H2
Canada
www.arclerpress.com
Email: orders@arclereducation.com

ISBN: 978-1-77469-503-6 (Hardcover)

ABOUT THE AUTHOR

Nelson Bolivar is currently a Physics Professor in the Physics Department at the Universidad Central de Venezuela, where he has been teaching since 2007. His interests include quantum field theory applied in condensed matter. He obtained his PhD in physics from the Universite de Lorraine (France) in 2014 in a joint PhD with the Universidad Central de Venezuela. His BSc in physics is from the Universidad Central de Venezuela.

TABLE OF CONTENTS

LIST OF FIGURES

LIST OF TABLES

LIST OF ABBREVIATIONS

AGN	active galactic nuclei
BE	binding energy
CDWs	charge-density waves
CERN	European Center for Nuclear Research
DVDs	digital versatile discs
FIB	focused-ion-beam
GRBs	gamma-ray bursts
GUT	grand unified theory
HST	Hubble Space Telescope
KE	kinetic energy
MBE	molecular beam epitaxy
MOVPE	metal-organic vapor phase epitaxy
PE	potential energy
PWR	pressure water reactor
QCD	quantum chromodynamics
QED	quantum electrodynamics
SHM	simple harmonic motion
SLAC	Stanford Linear Accelerator Center

PREFACE

Although Aristotle and Eratosthenes performed measurements and computations that led to today's physics, the field of physics has its modern origins in the work of Galileo, Newton, and others during the scientific revolution of the 16th and 17th centuries. Knowledge and practice of physics grew consistently over 200 to 300 years before another revolution in physics, which is the topic of this book. Classical physics, which was mostly created before 1895, is distinguished by some physicists from modern physics, which is based on discoveries made after 1895. The exact year is unimportant, but significant developments in physics occurred about 1900. During the lengthy reign of Queen Victoria of England, from 1837 to 1901, significant changes happened in the social, political, and intellectual spheres, but perhaps none were as significant as the astonishing advances made in physics. Maxwell's description and predictions of electromagnetic, for instance, contributed to the rapid development of modern communications. During this time period, thermodynamics also developed into an exact science. However, none of these accomplishments have had an impact of the discoveries and applications of contemporary physics in the 20th century. Never again would the world be the same.

Developments in Modern Physics is an introduction to the fundamental principles and domains of modern physics. The primary objective of the book is to help prepare engineering students for the higher division courses, as well as to provide physics majors and engineering students with an up-to-date account of modern physics.

Additionally, this book gives a comprehensive examination of foundational theory and experimentation. Appropriate for second-year undergraduate science and engineering students, this text provides a comprehensive introduction to the fundamental concepts and methods of modern physics, such as calculations of relativity, particle physics, quantum physics, high energy physics, and cosmology. A balanced instructional approach explores key concepts first from a historical viewpoint, then through a contemporary lens using pertinent experimental evidence and a discussion of recent advancements in the area. The emphasis on the correlation between principles and practices gives continuity and creates an easily-followable "storyline" for students.

The book has been divided into eight chapters. Each chapter introduces a particular topic related to modern physics. The topics covered in the book include Spacetime and General Relativity, Quantum Physics, Nuclear Physics, Elementary Particle Physics, Cosmology, and Modern Astrophysics, Physics of Semiconducting Lasers, Modern Physics, and Its Implications in Behavioral Sciences, etc.

Extensive educational tools facilitate comprehension by motivating students to reflect and enhancing their capacity to apply conceptual information to practical situations. Numerous exercises and actual examples are provided to reinforce essential concepts. The text is equally beneficial for students of physics and other multidisciplinary fields.

—***Author***

CHAPTER 1

INTRODUCTION TO MODERN PHYSICS

CONTENTS

1.1. INTRODUCTION

Physics is a natural science that uses experiments, measurements, and mathematical analysis to explain phenomena. Its goal is to discover quantitative physical principles for everything from the nanoscale to the solar system, to the cosmological scales. Physics is a field of science that in its core studies the interplay of energy and matter. Physics is often divided into two branches: modern and classical physics (Richtmyer et al., 1955).

Modern physics is the branch of physics that studies post-Newtonian principles in various fields. It is characterized by two significant breakthroughs in the 20th century: Quantum Mechanics and Relativity. Modern physics is primarily concerned with a sophisticated explanation of nature using ideas that arrives from classical physics. Most breakthroughs are based on quantum mechanics and Einstein's theory of relativity. Albert Einstein is often referred to as the "Father of Modern Physics" (Gil and Solbes, 1993).

Modern physics is a field of physics that uses post-Newtonian notions to study the basic origin of the world. Certain experimental discoveries in the early 20th century did not meet the expectations of classical physics, which examines physical phenomena on a small scale. Such hypotheses eventually gave birth to modern physics. Quantum theory explains small-scale physics and gravity, while relativity theory explains gravity and large-scale physics. Both theories may be used to approximate the outcomes of the classical theory (Brehm and Mullins, 1989).

Modern physics is concerned with providing the fundamentals and boundaries of contemporary physics. It concentrates on fields such as quantum mechanics, and applications in nuclear, atomic, particle, and condensed-matter physics, and also condensed-matter physics. Special relativity, relativistic quantum mechanics, Dirac equation, and Feynman diagrams, as well as quantum fields in conjunction with special relativity, are the primary topics covered. Modern physics aspires to provide a thorough treatment of such issues in appropriate depth (Jewett and Serway, 2008).

1.2. JOURNEY OF PHYSICS TO MODERN PHYSICS

1.2.1. Physics of Ancient Greece

In the past, people used to interpret each natural phenomenon with religious, supernatural, or mythological interpretations before the archaic period in

Greek history. It was the prevalent thinking before Thales of Miletus changed it (Figure 1.1) (Lang, 1999).

Figure 1.1. The architecture of Ancient Greece reflects its past.

Source: https://www.britannica.com/place/ancient-Greece.

Thales of Miletus, sometimes known as "the Father of Science," was a Greek astronomer and mathematician who first asserted that each phenomenon had a natural explanation. He also said that water is the fundamental component of all matter. Then Anaximander refuted Thales' theory, claiming that the fundamental element is a material called Apeiron (Einstein, 1934). Others like Parmenides, Heraclitus, Zeno of Elea, Empedocles, and Democritus proceeded in their footsteps. They developed Pre-Socratic philosophy, an ancient Greek philosophy that predated Socrates and had been unaffected by him (Rihll and Wilson, 2003).

The emergence of the idea of atomism, initially proposed by Leucippus and his disciple Democritus, was among the major achievements of this time. They explored the theory that all stuff in the cosmos is made up of several indestructible, indivisible components known as atoms.

A brilliant philosopher from Greece's ancient period made his imprint on history. He had been known as Aristotle, and he was the one who first recognized the value of observation as a means of uncovering the principles that govern natural occurrences (Hesse, 2005).

In the fourth century BCE, Aristotle authored the 1st treatise referring to that field of study as "Physics." He also developed the 4-element theory and attempted to describe the principles of gravity and motion. Archimedes created the ideas of equilibrium states and gravity centers after a long period. Such theories will later impact prominent academics such as Newton and Galileo (Lang, 2004).

The title "Father of Physics" is bestowed on three scientists from various eras, based on their most significant achievements (Pappas and Karas, 1987).

For his contributions to Astrophysics, Galileo Galilei is known as the Father of Observational Physics. The laws of gravity and motion were created by Sir Isaac Newton. His hypothesis underpins classical physics, and it works well on a small scale. He also taught calculus theory in mathematics. Newton is renowned as the Father of Physics because of his significant contributions (Watson, 1954).

Albert Einstein is widely regarded as the founder of modern physics. He established both the general and special theories of relativity. Such theories explain how things behave at great speeds (near the speed of light) and in the presence of gravity. For his description of the photoelectric effect, he received the Nobel Prize (Frantzeskakis, 2022).

1.2.2. Contributions from the Islamic World

Between the seventh and the 15th centuries in the Islamic world, a significant scientific revolution had been taking place.

With the widespread translation of Greek and Indian scientists' publications into Arabic, science became accessible to Islamic geniuses, allowing them to contribute to humanity's scientific history (Faruqi, 2006).

Ibn Al-Haytham was a prominent Arabic scholar during this period. He contributed significantly to science. Owing to his technique, which had been dependent upon experimental evidence and the repeatability of its conclusions, Ibn Al-Haytham was dubbed "the father of the modern scientific method" (Figure 1.2) (Hilgendorf, 2003).

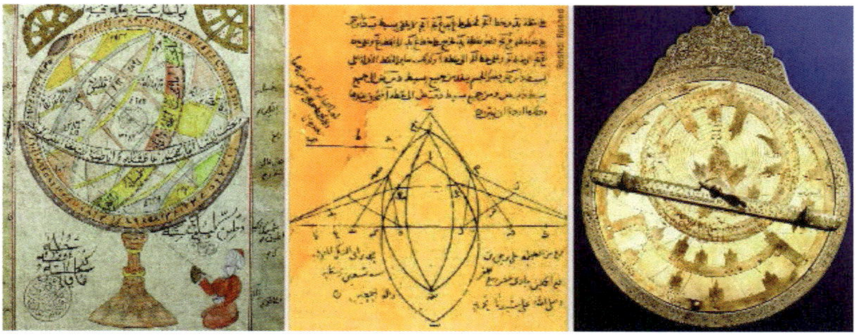

Figure 1.2. Arabic intellectuals and their works of art.

Source: https://physicsworld.com/a/science-in-the-muslim-world/.

Ibn Al-Haytham is known as the "Father of Optics" in physics. Light flows to the eye in beams from various locations on an object, as he proposed.

There had been Ibn Sina, a famous genius who dedicated to science in his book "Book of Healing." He described Newton's 1st law of motion inertia, which says that an item in motion would remain in motion until pushed on by an external entity (Hoodbhoy, 2007).

Another Islamic thinker, Abu'l-Barakat, addressed how a falling body accelerates as its momentum increases. And Ibn Bajjah often referred to as "Avempace" in Europe, demonstrated that every opposing force has a response force. He did not, however, mention that such forces are comparable. This was a precursor to Newton's laws of motion, which asserts that it is an opposite and equal response to each action (Hasan and Daud, 2021).

1.2.3. The Scientific Revolution

Copernicanism and the fight between mechanics and astronomy kick off the tale of the scientific revolution. The notion of natural motion and location was eliminated, and the circular motion of the Earth was inconsistent having Aristotelian physics. He made compelling arguments for the heliocentric model of the Solar System, presumably to improve the accuracy and ease of generation of planetary motion tables (Figure 1.3) (Bouchaud, 2008).

Figure 1.3. Galileo Galilei's contribution to scientific evolution.

Source: https://astronomy.com/news/2021/11/12-fascinating-facts-about-galileo-galilei.

The second round of the conflict began with Galileo Galilei, an Italian astronomer, philosopher, and mathematician who was the first to enter the arena. In particular, he is credited with making notable contributions to the disciplines of astronomy and motion, as well as to the creation of the scientific process. He was the first person to find four of Jupiter's Moon about 400 years before. In addition, he discovered the rule of free fall and the parabolic route of projectile motion, both of which are now known (Bomberg et al., 2020).

In the world of physics, another star began to shine as a result of a series of events. He brought the three laws of motion with him. Each of these laws defined the relationship between motion and the objects it explained. In addition, he was the first to propose the formula for universal gravity. Newton, the illustrious scholar, was the headliner of the show (Büttner et al., 2003).

The application of the universal gravitation law to the description of the motion of planets required the development of an entirely new field of mathematics, known as calculus. Newton's discovery of calculus was among his most important scientific accomplishments (Bechler, 2012).

Newton made several other significant discoveries, including the construction of the world's 1st functional telescope and the development of a theory of the colors dependent upon the observation using prism to decomposes white light into the various colors that make up the visual spectrum (Kirilyuk, 2008).

When he was younger, he explored the speed of sound and showed the generalized binomial theorem, as well as developed a method for estimating the roots of a function, among other things. Simon Stevin's decimals served as an inspiration for his work on infinite series (Frankel, 1976). In addition, by establishing the compatibility between Kepler's laws of planetary motion as well as his theory of gravitation, Newton was able to dispel the final remaining questions regarding heliocentrism that had been raised.

As a result of Newton's contributions, the scientific world was poised to usher in an entirely modern period of physics: the age of modern physics had arrived (Koyré, 1943).

1.2.4. The Birth of Modern Physics

Despite its successes at the end of the 19th century, classical physics confronted many constraints and significant problems that might not be handled by utilizing the rules of physics at the time. Classical physics' failure to describe certain physical processes, like the energy distribution in blackbody radiation and the photoelectric effect, is one example of such limits (Jones, 2007).

1.2.4.1. Radiation Experiments

Scientists began to identify novel kinds of radiation in the 19th century, like radioactive elements identified by Marie and Pierre Curie, and the electron identified by J. J. Thomson, X-rays identified by Wilhelm Rontgen.

Such findings led scientists to question the indestructibility of the atom and the nature of matter (Richtmyer et al., 1955). The classical theory also failed to describe the Michelson–Morley experiment, which demonstrated that there did not even appear to be a preferable reference for explaining electromagnetic events, at least regarding the imagined luminiferous ether. It also failed to describe radiation and radioactive decay until Lise Meitner and Otto Frisch identified nuclear fission, paving the way for the practical use of "atomic" energy (Podgoršak, 2014).

Modern physics' inception and development was a watershed moment in human history. It is because contemporary physics' principal ideas changed our understanding of the cosmos and sparked a massive scientific revolution.

Modern physics is a discipline of physics that deals with post-Newtonian notions in physics. It is dependent upon two significant 20th-century breakthroughs: quantum and relativity theory (Manikopoulos and Aquirre, 1977).

The word "modern physics" refers to current physics. This name alludes to the breakthrough that occurred following Newton's laws, thermodynamics, and Maxwell's equations which are all considered "classical" physics (Figure 1.4) (Baggott et al., 2004).

Figure 1.4. Modern physics visualizations.

Source: https://byjus.com/physics/modern-physics/.

As a result, modern physics may be thought of as among the latest step in the history of physics. This history may be traced back to ancient China, medieval Europe, the Islamic world, ancient Greece, and ancient India among other places. Later, the scientific revolution arose, which was dependent upon the theories of Galileo Galilei, Nicolaus Copernicus, Isaac Newton, René Descartes, and other luminaries of the time (Dardashti et al., 2019).

We will examine the history of physics during these periods as well as the development of the major theories of modern physics.

A brief discussion of the two most significant advances in contemporary physics that occurred in the early 20th century: quantum and relativity physics would be included in this chapter as well (Xu et al., 2014).

1.2.5. Albert Einstein and Relativity

It is 1905, Albert Einstein, a 26-year-old German physicist, is about to make a monumental breakthrough in the history of physics with the development of relativity theory. He claimed that motion between an observer and what is being viewed affects space and time measurements (Einstein, 1982).

Einstein's theory of relativity is regarded as the best intellectual achievement of all time. Einstein also stated that the speed of light in a vacuum is constant, that is, the same for all observers, and serves as an absolute universal physical quantity.

Additionally, he deduced the famous equation $E = mc^2$, which describes the mass-energy equivalency (Lorentz et al., 1952).

1.2.5.1. Special Relativity

The link between physical data and the ideas of time and space is described by the special theory of relativity. The inconsistencies between electromagnetic and Newtonian mechanics gave rise to this hypothesis, which resulted in significant progress in both fields (Sexl and Urbantke, 2012).

The fundamental historical question was whether it was worthwhile to analyze how electromagnetic waves travel in the presumed medium "ether" and how it moves with other objects. In his special theory of relativity, Einstein demolished the idea of "ether."

His core concept, on the other hand, does not include extensive electromagnetic theory (Alstein et al., 2021). The theory of special relativity attempted to address the enigma of "what is time?" "Absolute, real, and mathematical time, by itself, and from its nature, flowing equitably without regard to anything external, and by another name is termed duration," Newton said in the Principia (1686). All classic physics is dependent upon the definition (Figure 1.5) (Amelino-Camelia, 2002).

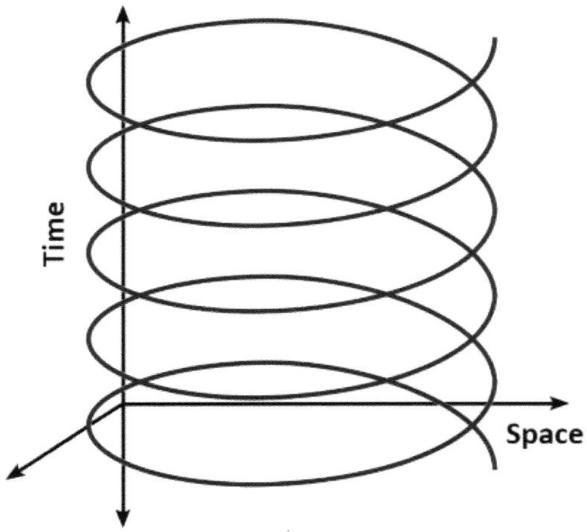

Figure 1.5. The Earth's worldline as it rounds the Sun. Although the Earth's route through space is almost round, it is always moving ahead in time.

Source: http://www.thestargarden.co.uk/Special-relativity.html.

This response, Einstein discovered, was insufficient. He provided his perspective (Mansouri and Sexl, 1977).

1.2.5.2. Relativity

As per Einstein, every "observer" must utilize his time frame, and the time scales of two observers in relative motion may vary. This influences position measurements as well. Time and space become inextricably linked notions, reliant on the observer. Every observer is in charge of his coordinate system or time-space framework (Figure 1.6) (Peres and Terno, 2004).

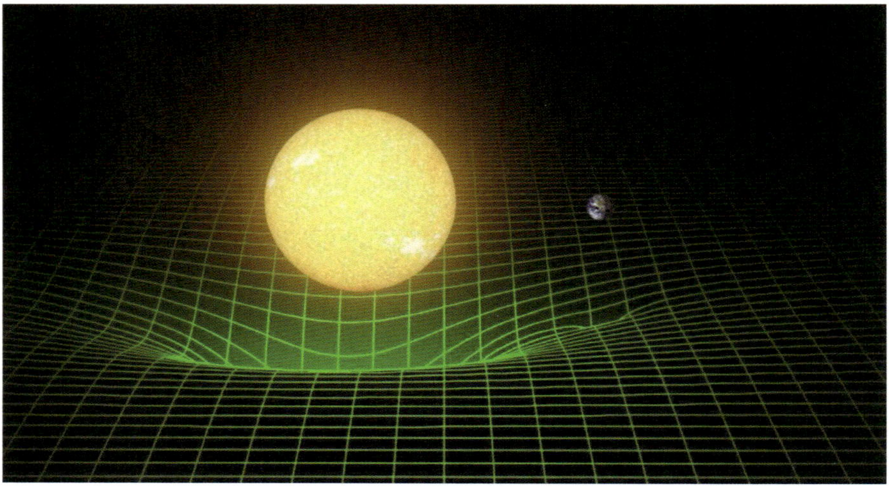

Figure 1.6. Graphical depiction of time and space.

Source: https://www.wonderopolis.org/wonder/what-is-the-space-time-continuum.

Einstein delved more into the nature of motion in the cosmos in 1916. He proposed the notion of spacetime curvature as a source of gravity, which later became the universal theory of relativity.

The typical perception of gravitational pull, according to Einstein, is a form of illusion generated by the geometry of space.

The mass of the item generates a curvature of spacetime around it, which defines the spacetime route that other freely moving objects must take. Newton's universal law of gravitation was fully supplanted by this new understanding of how gravity operates (Nottale, 1992).

The concepts of relativity are:

- No massive object may travel to a velocity equal or faster than the speed of light;
- For all observers, the rules of physics remain unchangeable;
- When an object approaches the speed of light, its length decreases, (length contraction). A ticking clock slows (time dilation);
- The masses of gravity and inertia are the same.

1.2.6. Quantum Physics

Another recent physics accomplishment attempted to examine a different realm, that of atoms and subatomic particles.

The issue of black body radiation was answered by the modern theory of quantum mechanics, which revealed that at shorter wavelengths, near the UV end of the spectrum, the energy approached zero, when classical theory anticipated it must grow unlimited (Figure 1.7) (Busch et al., 1997).

Figure 1.7. Quantum physics as a graphical depiction.

Source: https://www.livescience.com/33816-quantum-mechanics-explanation.html.

The theory that examines and models' atoms and subatomic systems is known as quantum mechanics. The theory was developed over the first 30 years of the 20th century.

Max Planck proposed the fundamental concepts of quantum theory in 1900 (Piron, 1976). The Compton Effect proved that light has momentum and may scatter off particles, Louis de Broglie claimed that matter may be understood as acting as a wave similarly as electromagnetic waves act like particles (wave-particle duality) (Zeilinger, 1999).

1.2.6.1. Black Body Radiation

The purpose of this experiment was to use the heated filament method to examine black body radiation and confirm Wien's law and the inverse square law.

The wavelength of the emitted radiation, which is dependent on the blackbody's temperature, affects the intensity of the radiation emitted (Dicke et al., 1965). In addition, the amount of radiation released is proportional to the square of the distance from the dark body.

Researchers may prove the 4th power law of radiation, which generates Planck's curves at various temperatures, and the inverse square law for EM radiation, by measuring the emitted radiation from a heated filament as a function of the filament's temperature, the wavelength of the emitted radiation, and distance from the black body (Figure 1.8) (Gallagher and Cooke, 1979).

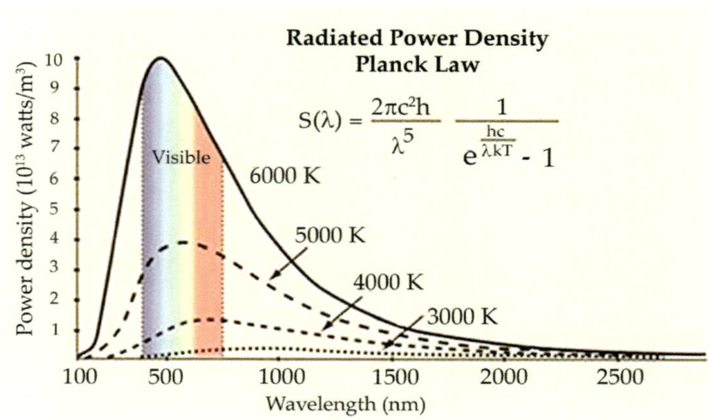

Figure 1.8. Radiation from black bodies.

Source: https://esfsciencenew.wordpress.com/2013/10/29/black-body-radiation/.

1.2.6.2. Laser Beam Divergence

This experiment confirmed that a laser beam's profile is Gaussian calculating its properties utilizing the laser photodiode technique.

In the transverse direction, many low-intensity laser resources output a laser beam with a Gaussian distribution $I(r) = Io_e - r_2/z_2$. Where z_2 is the diameter of the beam at which the intensity drops to Io/e_2 (Wright et al., 1992).

Also, because of the laser's coherent feature, it must not follow the inverse square law that regular light does (Figure 1.9).

Figure 1.9. Laser beam graphical depiction.

Source: https://www.vectorstock.com/royalty-free-vector/green-laser-beam-vector-30065757.

We acquire the profile of the laser beam, which must be Gaussian, by measuring the intensity of the laser beam as a function of the distance from the beam's center in the transverse direction utilizing a photodiode sensor. As a result, the diameter of the beam might be determined. The beam profile may be plotted at various distances from the resource to measure the beam divergence, demonstrating that the laser does not obey the inverse square law (Riley and Gusinow, 1977).

1.2.6.3. Laser Electro-Optic Effect

The purpose of this experiment was to investigate the electro-optic properties of many crystals by utilizing Kerr Cell Procedure.

A single wavelength of monochromatic polarized light (laser) is incident on a Lithium niobate crystal which is angled 45° to the vertical. When an electric field is applied to a crystal, it becomes birefringent. It is found that the phase shift between normal and unusual light is proportional to the square of the electric field (Van Exter et al., 1997).

A laser beam polarized to 45° with the vertical is used to light a lithium niobate crystal. An electric field is given to the crystal, allowing light to escape, and be recorded by the photosensor. A photosensor is used to record the quantity of light traveling through the crystal as a function of the electric field, and the half voltage value has been computed (Consoli et al., 2016).

1.2.6.4. Michelson' Interferometer

The purpose of this experiment was to use Michelson's interferometer to estimate the refractive index of a thin clear plate.

By splitting a monochromatic light beam from a laser resource into two beams, two mirrors reflect the two beams to a display, where an interference pattern is visible (Ikram and Hussain, 1999). By adjusting either one or more of the two mirrors, the phase distinction between the two beams varies, and the quantity of fringes across the field of vision changes proportionately.

The quantity of fringes crossing the field of view is tallied as either one (or even both) of the two mirrors is moved or the glass plate stage is rotated via angle. Thus, the laser's wavelength and the glass plate's refractive index might be measured (Figure 1.10) (Kersey et al., 1991).

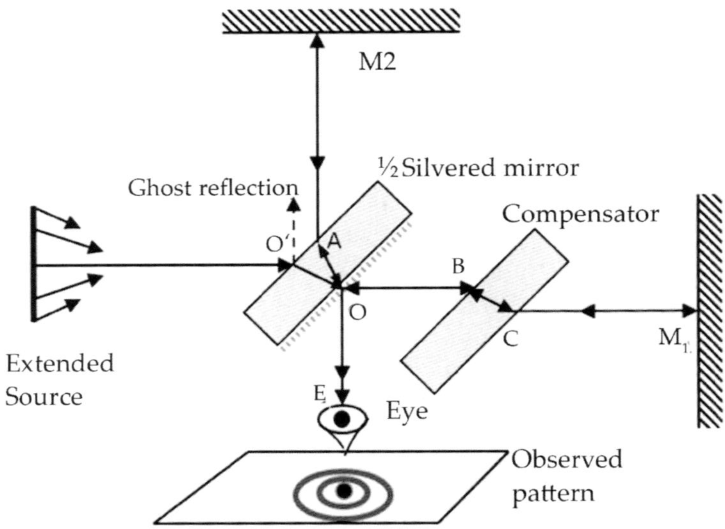

Figure 1.10. The interferometer of Michelson.

Source: http://www.physicsbootcamp.org/section-michelson-interferometer.html.

1.2.6.5. Millikan Oil Drop

The purpose of this experiment was to see if the Oil Drop Method can accurately quantify electric charge.

Oil droplets are sprayed across two plates in the presence of an electric field. An ionizing source gives the oil droplets a considerable charge (Jones, 1995). Consequently, the mass of the oil drops and the quantity of charge it has gained from the ionizing radiation impact its mobility between the plates. The intensity of the applied electric field and its polarity determine the charge's mobility, so it can rise, fall, or even remain fixed between the plates (Allen and Raabe, 1982).

We may calculate the amount of charge the oil droplets have collected by monitoring their fall and rise speeds in the existence of an electric field for oil drops. As a result, the amount of charge carried by every drop may be proven to be an integral multiple of the electron charge (Halyo et al., 2000).

1.3. IMPORTANT DISCOVERIES IN MODERN PHYSICS

Numerous experiments have shaped Modern Physics' history and growth. Those who contributed to a better knowledge of the structure of atoms and matter come to mind. The following are certain instances of significant discoveries (Bridgman et al., 1927):

- Wilhelm Röntgen discovered the presence of X-rays in 1895. It is a form of invisible, incredibly penetrating radiation.
- In the year 1900, German physicist Max Planck claimed that energy is charged by electromagnetic fields and has quantized quantities. It is the integral multiple of a small, fixed quantity.
- Albert Einstein described and demonstrated that references move at very higher speeds via his theory of relativity in 1905. That speed was near to the speed of light propagation, allowing you to experience distance and time measurement in a variety of ways.
- Niels Bohr suggested the quantization of the energy levels of electrons distributed around atomic nuclei in 1913. It indicates that their energy is proportional to an integer multiple of a minimal value.
- In 1924, scientist Louis De'Broglie created the wave-particle duality, which demonstrated that anybody may act like a wave.

- Quantum Mechanics was first published in 1926. It was created by physicists like Werner Heisenberg and Erwin Schrödinger.

1.4. ORIGIN OF THE THEORY OF RELATIVITY

Einstein recognized that time and space are the linked together in a single structure. Every observation and measurement are based on a frame of reference, space, and time are relative depending on the frame. In Newtonian physics, time is treated as a constant that is independent of the observer, all observers measure the same time which flows equally everywhere. Einstein proposed the concept of "spacetime." Gravity is the result of objects wrapping the fabric of spacetime. Einstein also discovered that the ideas of mass and energy are interchangeable (Holton, 1960).

1.5. SOLID STATE PHYSICS

Solid-state physics, crystallography, quantum mechanics, electromagnetism, and metallurgy are all utilized to investigate rigid matter or solids. In condensed matter physics, this is the most significant sub-discipline. Solid-state physics studies how large-scale properties of materials are obtained from their atomic-scale properties. Consequently, the theoretical underpinning of materials research in solid-state physics. It also has direct uses, such as in transistor and semiconductor technologies (Jones and March, 1985).

The majority of solid-state physics is based on the theory of crystals. This is mostly because the periodicity of atoms in a crystal is a distinctive trait that facilitates mathematical modeling. Likewise, crystalline materials often include magnetic, electrical, mechanical, or optical properties that may be utilized in engineering (Ibach and Lüth, 2009).

1.6. ATOMIC THEORY

Atomic theory is the scientific notion that matter is composed of microscopic constituents called atoms. The beginnings of atomic theory can be traced back to atomism, an old philosophical tradition. As per this idea, slicing a lump of material into smaller and smaller pieces will eventually result in the portions being unable to be sliced into smaller pieces. Ancient Greek philosophers named these hypothesized essential constituents of matter atoms, which mean "uncutable" (Bohr, 1925).

1.7. JOHN DALTON

John Dalton explored and improved on this prior work, proposing a new notion eventually recognized as the rule of many proportions: if the similar two elements may be mixed to make multiple compounds, the ratios of the two elements' masses in each compound would be expressed by tiny whole numbers. It was a common trend in chemical reactions that Dalton and other scientists noticed at the time (Dalton, 2010).

1.8. AVOGADRO

In 1811, Amedeo Avogadro resolved the flaw in Dalton's theory in principle. As per Avogadro, equivalent quantities of any two gases, at the same pressure and temperature, have the same number of molecules (in other words, the particle mass of gas does not influence the volume it fills) (Hanwell et al., 2012). Avogadro's law allowed him to deduce the diatomic nature of several gases by studying the volumes at which gases interacted. A single oxygen molecule breaks in half to form two water particles when 2 liters of hydrogen react with one liter of oxygen to produce 2 liters of water vapor (at constant pressure and temperature). As a consequence, Avogadro was capable to calculate the atomic mass of oxygen and other elements with greater precision, and also differentiating between atoms and molecules (Becker, 2001).

1.9. BROWNIAN MOTION

In 1827, a British botanist named Robert Brown noted that dust particles among pollen grains floating in water jiggled around for no obvious cause. In 1905, Albert Einstein claimed that Brownian motion is caused via water molecules continually moving grains around, and he built a hypothetical mathematical model to describe it. In 1908, French physicist Jean Perrin experimentally proved this concept, providing further evidence for particle theory (Mörters and Peres, 2010).

As a result, post-Newtonian notions are used in modern physics to deal with the underlying nature of the cosmos. Quantum theory and relativity theory are two foundations of contemporary physics. Three scientists have been dubbed "Father of Physics" for their contributions to Astrophysics in various eras (Uhlenbeck and Ornstein, 1930).

1.10. THE ADVENT OF QUANTUM THEORY

The experimental data of the photoelectric effect, black body radiation, and electron interference phenomena, as well as the stability of an atom, were not explained by classical physics. Waves and particles are treated differently in classical physics. Max Planck proposed in 1900 that light is made up of packets or quanta of energy called photons. Every photon possesses energy (Van Vleck and Sherman, 1935):

$$E = hv$$

The frequency of light is v, while Planck's constant is h.

This idea might describe the black body radiation phenomena, even though it violates the traditional theory that views light to be an electromagnetic wave. Einstein correctly described the photoelectric phenomenon in 1905 by imagining light as a swarm of photons (quanta of energy).

The interference of electrons and the stability of an atom, on either hand, might only be represented if electrons were thought of as waves. Each particle, according to De Broglie, acts like a wave with a wavelength of (Claverie and Diner, 1980):

$$\lambda = hp$$

The momentum is represented by p. The wavelengths of everyday items are so short that classical theory works on a small scale, yet the wavelengths of subatomic particles such as electrons are equivalent to their size (Holevo, 2003).

Quantum theory was found to be essential to describe physics at tiny sizes (for example, the atomic scale). Energy, angular momentum, and other characteristics of a bound system are quantized in this theory. From a mathematical standpoint, several scientists such as Bohr, Heisenberg, Schrödinger, Pauli, and Dirac proposed the theory. Quantum Field Theory evolved in the late 20th century as a result of the work of scientists such as Jordan, Hawking, Weinberg, and Feynman (Ashtekar and Lewandowski, 1997). The fundamental notions of quantum theory are discussed in subsections.

1.10.1. Wave-Particle Duality

Light is both a particle and a wave. Photons, or energy quanta, make up light. The nature of particles is wavelike. In space, particles are delocalized (Arndt et al., 1999).

1.10.2. Uncertainty Principle

It is impossible to measure a particle's accurate position and momentum at the same time (Busch et al., 2007).

1.10.3. Measurement Problem

Measuring or viewing a system alters its condition (Wigner, 1963).

REFERENCES

1. Allen, M. D., & Raabe, O. G., (1982). Re-evaluation of Millikan's oil drop data for the motion of small particles in the air. *Journal of Aerosol Science, 13*(6), 537–547.
2. Alstein, P., Krijtenburg-Lewerissa, K., & Van, J. W. R., (2021). Teaching and learning special relativity theory in secondary and lower undergraduate education: A literature review. *Physical Review Physics Education Research, 17*(2), 023101.
3. Amelino-Camelia, G., (2002). Doubly-special relativity: First results and key open problems. *International Journal of Modern Physics D, 11*(10), 1643–1669.
4. Arndt, M., Nairz, O., Vos-Andreae, J., Keller, C., Van, D. Z. G., & Zeilinger, A., (1999). Wave-particle duality of C60 molecules. *Nature, 401*(6754), 680–682.
5. Ashtekar, A., & Lewandowski, J., (1997). Quantum theory of geometry: I. Area operators. *Classical and Quantum Gravity, 14*(1A), 55.
6. Baggott, J. E., Baggott, J., & Baggott, E. O. L. M. J., (2004). *Beyond Measure: Modern Physics, Philosophy, and the Meaning of Quantum Theory* (Vol. 1, pp. 1–7). Oxford University Press on Demand.
7. Bechler, Z., (2012). *Newton's Physics and the Conceptual Structure of the Scientific Revolution* (Vol. 127, pp. 2–9). Springer Science & Business Media.
8. Becker, P., (2001). History and progress in the accurate determination of the Avogadro constant. *Reports on Progress in Physics, 64*(12), 1945.
9. Bohr, N., (1925). Atomic theory and mechanics. *Nature, 116*(2927), 845–852.
10. Bomberg, M., Romanska-Zapala, A., & Yarbrough, D., (2020). Journey of American Building Physics: Steps leading to the current scientific revolution. *Energies, 13*(5), 1027.
11. Bouchaud, J. P., (2008). Economics needs a scientific revolution. *Nature, 455*(7217), 1181.
12. Brehm, J. J., & Mullins, W. J., (1989). *Introduction to the Structure of Matter: A Course in Modern Physics, 1*, 960.
13. Bridgman, P. W., Bridgman, P. W., Bridgman, P. W., & Bridgman, P. W., (1927). *The Logic of Modern Physics* (Vol. 3, pp. 1–5). New York: Macmillan.

14. Busch, P., Grabowski, M., & Lahti, P. J., (1997). *Operational Quantum Physics* (Vol. 31, pp. 1–6). Springer Science & Business Media.
15. Busch, P., Heinonen, T., & Lahti, P., (2007). Heisenberg's uncertainty principle. *Physics Reports, 452*(6), 155–176.
16. Büttner, J., Renn, J., & Schemmel, M., (2003). Exploring the limits of classical physics: Planck, Einstein, and the structure of a scientific revolution. *Studies in the History and Philosophy of Modern Physics, 34*(1), 37–59.
17. Claverie, P., & Diner, S., (1980). The concept of molecular structure in quantum theory: Interpretation problems. *Israel Journal of Chemistry, 19*(1–4), 54–81.
18. Consoli, F., De Angelis, R., Duvillaret, L., Andreoli, P. L., Cipriani, M., Cristofari, G., & Verona, C., (2016). Time-resolved absolute measurements by electro-optic effect of giant electromagnetic pulses due to laser-plasma interaction in nanosecond regime. *Scientific Reports, 6*(1), 1–8.
19. Dalton, J., (2010). *A New System of Chemical Philosophy* (Vol. 1, pp. 3–9). Cambridge University Press.
20. Dardashti, R., Hartmann, S., Thébault, K., & Winsberg, E., (2019). Hawking radiation and analog experiments: A Bayesian analysis. *Studies in History and Philosophy of Science Part B: Studies in History and Philosophy of Modern Physics, 67*(1), 1–11.
21. Dicke, R. H., Peebles, P. J. E., Roll, P. G., & Wilkinson, D. T., (1965). Cosmic blackbody radiation. *The Astrophysical Journal, 142*(1), 414–419.
22. Einstein, A., & Rosen, N., (1935). The particle problem in the general theory of relativity. *Physical Review, 48*(1), 73.
23. Einstein, A., (1934). On the method of theoretical physics. *Philosophy of Science, 1*(2), 163–169.
24. Einstein, A., (1982). How i created the theory of relativity. *Physics Today, 35*(8), 45–47.
25. Faruqi, Y. M., (2006). Contributions of Islamic scholars to the scientific enterprise. *International Education Journal, 7*(4), 391–399.
26. Frankel, E., (1976). Corpuscular optics and the wave theory of light: The science and politics of a revolution in physics. *Social Studies of Science, 6*(2), 141–184.

27. Frantzeskakis, L., (2022). *What Did the Greek Philosophers Contribute Towards the Atom Theory from Ancient Greece to Modern Times, 1*, 4–8.
28. Gallagher, T. F., & Cooke, W. E., (1979). Interactions of blackbody radiation with atoms. *Physical Review Letters, 42*(13), 835–840.
29. Gil, D., & Solbes, J., (1993). The introduction of modern physics: Overcoming a deformed vision of science. *International Journal of Science Education, 15*(3), 255–260.
30. Halyo, V., Kim, P., Lee, E. R., Lee, I. T., Loomba, D., & Perl, M. L., (2000). Search for free fractional electric charge elementary particles using an automated Millikan oil drop technique. *Physical Review Letters, 84*(12), 2576.
31. Hanwell, M. D., Curtis, D. E., Lonie, D. C., Vandermeersch, T., Zurek, E., & Hutchison, G. R., (2012). Avogadro: An advanced semantic chemical editor, visualization, and analysis platform. *Journal of Cheminformatics, 4*(1), 1–17.
32. Hasan, N. A. H., & Daud, M. A., (2021). Islamic civilization: Contributions and achievements. *Jurnal'Ulwan, 6*(1), 278–294.
33. Hesse, M. B., (2005). Forces and fields: The concept of action at a distance in the history of physics (Vol. 4, No. 1, pp. 2–7). Courier Corporation.
34. Hilgendorf, E., (2003). Islamic education: History and tendency. *Peabody Journal of Education, 78*(2), 63–75.
35. Holevo, A. S., (2003). *Statistical Structure of Quantum Theory* (Vol. 67, pp. 1–5). Springer Science & Business Media.
36. Holton, G., (1960). On the origins of the special theory of relativity. *American Journal of Physics, 28*(7), 627–636.
37. Hoodbhoy, P. A., (2007). Science and the Islamic world—the quest for rapprochement internal causes led to the decline of Islam's scientific greatness long before the era of mercantile imperialism. To contribute once again, Muslims must be introspective and ask what went wrong. *Science, 1*, 1–8.
38. Ibach, H., & Lüth, H., (2009). *Solid-State Physics: An Introduction to Principles of Materials Science* (Vol. 1, pp. 2–9). Springer Science & Business Media.
39. Ikram, M., & Hussain, G., (1999). Michelson interferometer for precision angle measurement. *Applied Optics, 38*(1), 113–120.

40. Jewett, J. W., & Serway, R., (2008). Physics for scientists and engineers with modern physics. *Vectors, 1*(2), 1–7.
41. Jones, D. G. C., (2007). The birth of modern physics. *Contemporary Physics, 48*(4), 227–229.
42. Jones, R. C., (1995). The Millikan oil-drop experiment: Making it worthwhile. *American Journal of Physics, 63*(11), 970–977.
43. Jones, W., & March, N. H., (1985). *Theoretical Solid-State Physics* (Vol. 35, pp. 1–7). Courier Corporation.
44. Kersey, A. D., Marrone, M. J., & Davis, M. A., (1991). Polarization-insensitive fiber optic Michelson interferometer. *Electronics Letters, 27*(6), 518–520.
45. Kirilyuk, A. P., (2008). The last scientific revolution. *Against the Tide: A Critical Review by Scientists of How Physics and Astronomy Get Done* (Vol. 1, p. 179).
46. Koyré, A., (1943). Galileo and the scientific revolution of the seventeenth century. *The Philosophical Review, 52*(4), 333–348.
47. Lang, S. B., (1999). The history of pyroelectricity: From ancient Greece to space missions. *Ferroelectrics, 230*(1), 99–108.
48. Lang, S. B., (2004). A 2400 year history of pyroelectricity: From ancient Greece to exploration of the solar system. *British Ceramic Transactions, 103*(2), 65–70.
49. Lorentz, H. A., Einstein, A., Minkowski, H., Weyl, H., & Sommerfeld, A., (1952). *The Principle of Relativity: A Collection of Original Memoirs on the Special and General Theory of Relativity* (Vol. 1, pp. 2–9). Courier Corporation.
50. Manikopoulos, C. N., & Aquirre, J. F., (1977). Determination of the blackbody radiation constant hc/k in the modern physics laboratory. *American Journal of Physics, 45*(6), 576–578.
51. Mansouri, R., & Sexl, R. U., (1977). A test theory of special relativity: I. Simultaneity and clock synchronization. *General relativity and Gravitation, 8*(7), 497–513.
52. Mörters, P., & Peres, Y., (2010). *Brownian Motion* (Vol. 30, pp. 4–7). Cambridge University Press.
53. Nottale, L., (1992). The theory of scale relativity. *International Journal of Modern Physics A, 7*(20), 4899–4936.

54. Pappas, V., & Karas, I., (1987). The printed book of physics: The dissemination of scientific thought in Greece 1750–1821 before the Greek revolution. *Annals of Science, 44*(3), 237–244.
55. Peres, A., & Terno, D. R., (2004). Quantum information and relativity theory. *Reviews of Modern Physics, 76*(1), 93–98.
56. Piron, C., (1976). On the foundations of quantum physics. In: *Quantum Mechanics, Determinism, Causality, and Particles* (Vol. 1, pp. 105–116). Springer, Dordrecht.
57. Podgoršak, E. B., (2014). Introduction to modern physics. In: *Compendium to Radiation Physics for Medical Physicists* (Vol. 1, pp. 1–115). Springer, Berlin, Heidelberg.
58. Richtmyer, F. K., Kennard, E. H., & Cooper, J. N., (1955). *Introduction to Modern Physics* (Vol. 747, pp. 2–6). New York: McGraw-Hill.
59. Rihll, T. E., & Wilson, A. G., (2003). Modeling settlement structures in ancient Greece: New approaches to the polis. In: *City and Country in the Ancient World* (Vol. 1, pp. 78–115). Routledge.
60. Riley, M. E., & Gusinow, M. A., (1977). Laser beam divergence utilizing a lateral shearing interferometer. *Applied Optics, 16*(10), 2753–2756.
61. Sexl, R. U., & Urbantke, H. K., (2012). *Relativity, Groups, Particles: Special Relativity and Relativistic Symmetry in Field and Particle Physics* (Vol. 7, pp. 1–5). Springer Science & Business Media.
62. Uhlenbeck, G. E., & Ornstein, L. S., (1930). On the theory of the Brownian motion. *Physical Review, 36*(5), 823.
63. Van, E. M. P., Van, D. A. J., & Woerdman, J. P., (1997). Electro-optic effect and birefringence in semiconductor vertical-cavity lasers. *Physical Review A, 56*(1), 84555–84558.
64. Van, V. J. H., & Sherman, A., (1935). The quantum theory of valence. *Reviews of Modern Physics, 7*(3), 167.
65. Watson, E. C., (1954). Reproductions of prints, drawings, and paintings of interest in the history of physics. 63. Conservation of momentum in ancient Greece. *American Journal of Physics, 22*(7), 477, 478.
66. Wigner, E. P., (1963). The problem of measurement. *American Journal of Physics, 31*(1), 6–15.

67. Wright, D., Greve, P., Fleischer, J., & Austin, L., (1992). Laser beam width, divergence and beam propagation factor—An international standardization approach. *Optical and Quantum Electronics, 24*(9), S993–S1000.

68. Xu, D., Liu, T., Li, H., Hua, J., Zhao, X., Tian, N., & Zhou, G., (2014). Ground-based platforms for space radiation research at the institute of modern physics. *Rendiconti Lincei, 25*(1), 13–16.

69. Zeilinger, A., (1999). Experiment and the foundations of quantum physics. In: *More Things in Heaven and Earth* (Vol. 1, pp. 482–498). Springer, New York, NY.

CHAPTER 2

SPACETIME AND GENERAL RELATIVITY

CONTENTS

2.1. INTRODUCTION

Hermann Minkowski (1864–1909), a former teacher of Einstein, began to geometrize relativity soon after the publication of the special theory of relativity. He asserted that space and time are inextricably linked. "Nobody has observed a position unless it is at the moment or a time unless it is at a place," he says. He names a world-point a point of space at a point of time, which is a system characterized by a set of numbers, x, y, z, and t (Ahsan, and Ali, 2017).

We could construct the space after ranging over all imaginable x, y, z, t values." To keep things simple, we will only look at one space dimension: the x-coordinate. Any occurrence in spacetime is referred to as an event, and it is depicted in Figure 2.1(a) (Geroch, 1968). Let us imagine the incident is the bursting of a firecracker. This event takes place at the world point, which has the x-coordinate and t-coordinate. (Several writers of highly developed relativity literature swap the coordinates, presenting the time-axis in the vertical direction to stress that this is not a standard depiction of distance vs. time.) Although the learner has been famous for the graphical format, we shall employ it in this book to make spacetime ideas simpler) (Crutchfield and Mitchell, 1995).

An illustration of a particle at rest in position x is depicted in Figure 2.1(b) as a world line. It is demonstrated by the graph that the particle is still traveling through time although it is at rest in space (Ahsan, 2005).

2.2. SPACETIME DIAGRAMS

However, while its x-coordinate remains constant because this is not traveling via space, its time-coordinate is constantly growing, indicating that it is moving via time. Figure 2.1(c) depicts a rod that is at rest in spacetime. The top line represents the world line of the rod's end at x_2, and the bottom line indicates the world line of the rod's opposite end at x_1. The bottom line indicates the world line of the rod's extreme side at x_1 (Dray, 2013; Hameroff, 2001). The bottom line denotes the world line of the rod's opposite end, which is located at point x_1. Take note of how the motion of the stationary rod sweeps a spacetime region with it. Illustrated in Figure 2.1(d) are the world lines of particle A, which travels at a constant velocity V_A, and particle B, which travels at a constant speed V_B. In the case of a particle, its velocity is represented by the slope of a straight line drawn on an x vs t graph depicting its position (Leubner, 1981). As the slope increases velocities will also increase. Because particle A has a steeper slope than particle B, particle A has a greater velocity, which is expressed as V_A greater than V_B. Because of the change in velocity of a particle over time, the world line of that particle is no straighter, but is instead curved, as seen in Figure 2.1(e). A particle being accelerated causes the world line of the particle to be bent in spacetime. The world-line of a mass is linked to a spring which is depicted in Figure 2.1(f). Note that the world line is bent throughout which indicates that this is a case of rapid acceleration (Ellis, 2006).

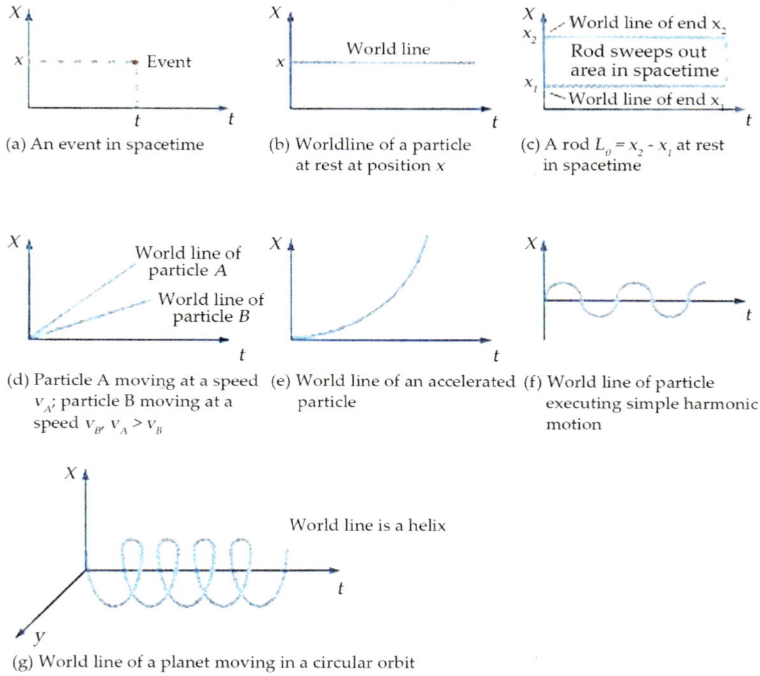

Figure 2.1. Spacetime diagrams.

Source: https://physics.stackexchange.com/questions/587399/why-are-the-co-ordinate-axes-of-a-moving-frame-k-tilted-in-spacetime-diagrams.

Figure 2.1(g) depicts a planet orbiting the sun in two dimensions. The planet moves in the x, y plane, but because it also moves in time, its world line deviates from the plane and becomes a helix. Consequently, when a planet departs from position x, completes one orbit, and comes back to the similar space point x, it is not in the identical spacetime position. It has advanced in time (During, 2012).

Changing the time axis to τ gives a more convenient depiction in spacetime diagrams, where;

$\tau = ct$ (1)

The τ is essentially a length in this form. (A length equals the multiplication of velocity times and the duration). The length τ is the distance traveled by light at a particular time. When t is converted to seconds then τ becomes a light second, which is the precise distance traveled by light in one second (Brill and Jacobson, 2006).

$\tau = \mathrm{ct} = (3.00 \times 10^8 \text{ m/s})(1.00\text{s}) = 3.00 \times 10^8 \text{ m}$

when; *t* is expressed in years, subsequently *τ* becomes a light-year, which is the distance traveled by light in one year.

$\tau = \mathrm{ct} = (3.00 \times 10^8 \text{ m/s})(1 \text{ yr})(\frac{365\,days}{1\,yr})\,(\frac{24\,hr}{1\,day})\,(\frac{3600s}{1\,hr})$
$= 9.47 \times 10^{15} \text{ m} = 9.47 \times 10^{12} \text{ km}$

The light-year is a common measurement of distance in astronomy. We create the spacetime diagram in Figure 2.2 using this new terminology. In this figure, a straight line may still indicate velocity (Janis, 2018).

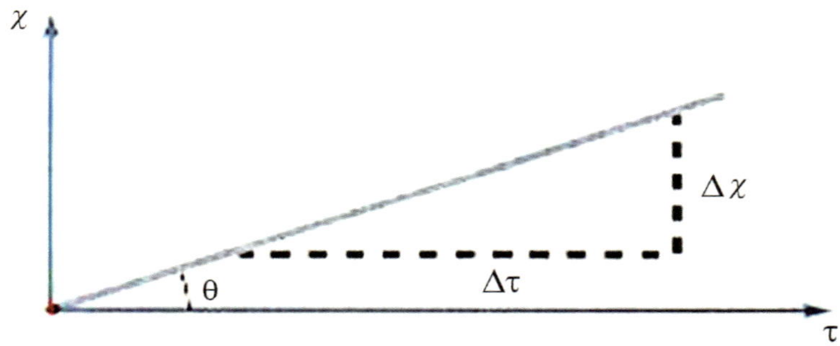

Figure 2.2. Changing the τ-axis to a τ-axis.

Source: https://www.researchgate.net/figure/Changes-in-spatial-QRS-T-angle-z-axis-Tax-and-corrected-QT-QTc-interval-calculated_fig1_346059134.

Nevertheless, because velocity is given as (Dikshit, 2013):

$$v = \frac{dx}{dt}$$

and while $\tau = ct$,

$cdt = d\tau$

or,

$$dt = \frac{d\hat{o}}{c}$$

Consequently, velocity becomes:

$$v = \frac{dx}{dt} = \frac{\frac{dx}{d\hat{o}}}{c} = \frac{cdx}{d\hat{o}}$$

However, $\frac{\Delta x}{\Delta \hat{o}}$ is the line's slope and which is provided by:

$$\frac{\Delta x}{\Delta \hat{o}} = \text{slope of line} = \tan\theta$$

On such a figure, the velocity is given by (Favaron et al., 2021):

$v = c \tan\theta$ (2)

If $\theta = 45°$ as a specific example in such a figure, the tan $45° = 1$ then Eqn. (2) becomes:

$v = c$

On the diagram of spacetime of x vs τ, a straight line at a 45° angle symbolizes the light signal's world line (Deshko, 2022).

On a spacetime diagram, we depict a light source at the origin emitting a beam of light concurrently to the right and the left, as illustrated in Figure 2.3. The world-line of a light ray emitted to the right is OL, while the world-line of a light ray emitted to the left is OL.' Because a particle's velocity should be less than c, any particle at O should have a world line having a slope of less than 45° and be enclosed inside the two light world lines OL and OL' (Witten, 2001).

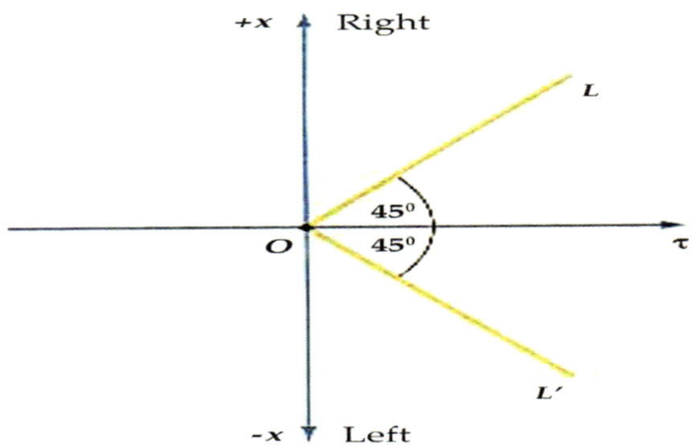

Figure 2.3. World lines of rays of light.

Source: https://www.researchgate.net/figure/Blue-lines-are-the-worldlines-of-four-particles-subluminal-from-the-point-of-view-of-O_fig2_356985390.

The τ-axis is the world line of the particle at O while it is at rest. The light cone illustrated in Figure 2.4 is obtained by extending the design in Figure 2.3 into two spatial dimensions. Straight lines traveling through O and confined inside the light cone might be the world lines of a particle or observer at the origin of O (Leonardi et al., 2022).

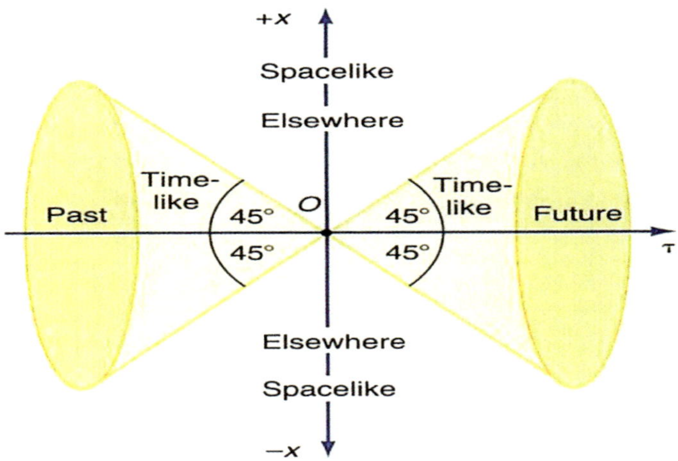

Figure 2.4. The light cone.

Source: https://plato.stanford.edu/index.html.

A world line included within the left-hand cone originates from the past of the observer, while any world lines contained within the right-hand cone originate from the future of the observer. Just world lines included contained inside the cone may have a causal influence on a particle or observer at O (Rousseaux et al., 2008). On the exterior of the cone; worldlines have no influence on the observer or particle at point O and are the worldlines of another observer or particle. Events that we essentially "see" are located on the light cone, as we perceive them via light beams. Since the world lines inside the cone are reachable in time, they are also referred to as time like. Events exterior of the cone is said to be spacelike as they originate in a space region that is unreachable to us, and thus referred to as elsewhere (Flores, 2007).

2.3. THE INVARIANT INTERVAL

Everything seems to be relative based on what has been mentioned so far. Is there anything else that remains unchanged in the shifting universe of spacetime? Is there anything on which all observers may agree independent of their state of motion? We have been continuously seeking certain motion constants in the realm of physics (Ahn et al., 2007). Remember how, in General Physics, we looked at the projectile motion of a particle in a single dimension and found that, while the bullet's location and velocity changed over time, one thing stayed constant: the projectile's total energy. Isn't there a constant of motion in spacetime, just like we ask? Yes, it is possible. The invariant interval is the constant value that all observers agree on irrespective of its position (Puig et al., 2003).

Let us look at the Lorentz transformation for the x-coordinate, which is given by the equation:

$$x' = \frac{x - vt}{\sqrt{1 - \frac{v^2}{c^c}}}$$

The differential Δx' becomes:

$$\Delta x' = \frac{dx - vdt}{\sqrt{1 - \frac{v^2}{c^c}}} \qquad (3)$$

Let us look at the Lorentz transformation for the t-coordinate, which is an equation (Chambon et al., 2016):

$$t' = \frac{t - xv/c^2}{\sqrt{1 - \frac{v^2}{c^c}}}$$

Taking the time differential Δt' we obtain:

$$\Delta t' = \frac{dt - dx\left(\frac{v}{c^2}\right)}{\sqrt{1 - \frac{v^2}{c^c}}} \qquad (4)$$

$$(\Delta x') = \frac{dt - dx\left(\frac{v}{c^2}\right)}{\sqrt{1 - \frac{v^2}{c^c}}}$$

To find the answer, we will square every one of such transformation equations (Wei, 1994):

$$(\Delta x')^2 = \frac{(dx)^2 - 2vdxdt + v^2(dt)^2}{1 - \frac{v^2}{c^c}} \qquad (5)$$

$$(\Delta t')^2 = \frac{(dt)^2 - 2vdxdt/c^2 + v^2/c^4(dx)^2}{1 - \frac{v^2}{c^c}} \qquad (6)$$

Suppose multiply Eqn. (6) by c^2 to obtain:

$$c^2(\Delta t')^2 = \frac{c^2(dt)^2 - 2vdxdt + v^2/c^2(dt)^2}{1 - \frac{v^2}{c^c}} \qquad (7)$$

Now subtract Eqns. (5) from (7) to obtain:

$$c^2(\Delta t')^2 - (\Delta x')^2 = \frac{c^2(dt)^2 - 2vdxdt + v^2/c^2(dx)^2}{1-\frac{v^2}{c^c}} - \frac{(dx)^2 - 2vdxdt + v^2(dt)^2}{1-\frac{v^2}{c^c}}$$

$$= \frac{c^2(dt)^2 + \left(\frac{v^2}{c^c}\right)(dx)^2 - (dx)^2}{1-\frac{v^2}{c^c}}$$

$$= \frac{(c^2 - v^2) + (dt)^2 - \left(1-\frac{v^2}{c^c}\right)(dx)^2}{1-\frac{v^2}{c^c}}$$

$$c^2 \ddot{A}t'^2 - \ddot{A}x'^2 = \frac{c^2\left(1-\frac{v^2}{c^c}\right) + (dt)^2 - \left(1-\frac{v^2}{c^c}\right)(dx)^2}{1-\frac{v^2}{c^c}} \qquad (8)$$

By dividing every term on the right by $1 - v^2/c^2$ we get:

$c^2(\Delta t')^2 - c^2(\Delta x')^2 = c^2(\Delta t)^2 + c^2(\Delta x)^2$ (9)

Eqn. (9) shows that the quantity $c^2(\Delta t)^2 - (\Delta x)^2$ as observed by observer S is identical to the amount $c^2(\Delta t')^2 - (\Delta x')^2$ as observed by observer S.' How can that be? This may only be true if both sides of Eqn. (9) equal a constant. Consequently, the amount $c^2(\Delta t)^2 - (\Delta x)^2$ is constant. Thus, it applies to all inertial systems. This amount is indicated by $(\Delta s)^2$ and is known as the invariant interval. Therefore, the interval of invariance is given by (Hu and Wang, 2000):

$(\Delta s)^2 = c^2(\Delta t)^2 - (\Delta x)^2$ (10)

As a result, the constant interval is a spacetime constant. This value in spacetime is agreed upon by all observers, irrespective of its state of motion. When the extra two spatial dimensions are added, the 4-dimensional constant interval becomes:

$(\Delta s)^2 = c^2(\Delta t)^2 - (\Delta x)^2 - (\Delta y)^2 - (\Delta z)^2$ (11)

We find the constant interval of spacetime to be an odd amount. Pythagorean theorem defines a constant interval in ordinary space, not spacetime, as (Almeida and Ledberg, 2010):

$(\Delta s)^2 = (\Delta x)^2 + (\Delta y)^2 = (\Delta x')^2 + (\Delta y')^2$ (12)

As seen in Figure 2.5, wherein; Δs are constant, it is clear that is not greater than the circle's radius provided by Eqn. (12) and depicted in picture 2.5.

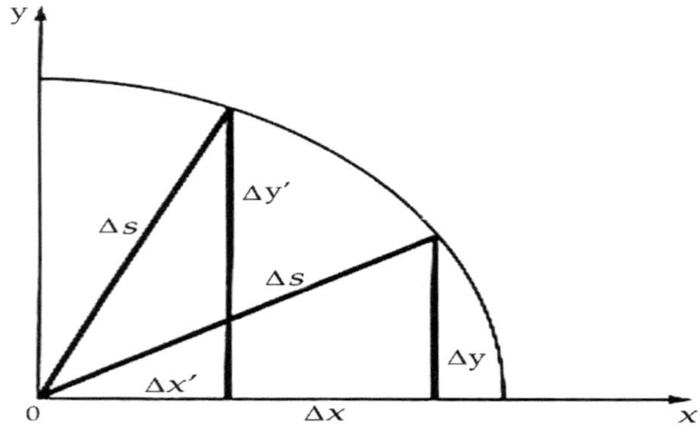

Figure 2.5. The invariant interval of space.

Source: http://www.herongyang.com/Physics/Minkowski-Diagram-Showing-Spacetime-Interval.html.

Eqn. (12) is, in other words, the circle equation $r^2 = x^2 + y^2$. Although Δx and Δx' differ, and Δy and Δy' differ, the quantity Δs have been a similar quantity that is positive (Bowen, 1979).

Let us examine Eqn. (10) for the spacetime constant interval. But firstly, as we have done in Eqn. (1), let ct = τ. The invariant interval, Eqn. (10), can thus be expressed as:

$$(\Delta s)^2 = (\Delta \tau)^2 - (\Delta x)^2 \qquad (13)$$

Since the negative sign in front of $(\Delta x)^2$, the equation is a hyperbola, $x^2 - y^2$ = constant, instead of the equation of a circle ($x^2 + y^2 = r^2$). The Pythagorean Theorem gives the distance between two points in Euclidean geometry, which has been expressed by the hypotenuse of a right-angle triangle: The hypotenuse's square is equivalent to the complete squares of the other two triangle sides (Valand, 1968). The square of the interval Δs in spacetime, on either hand, is equal to the difference of the squares of the other two sides, not the sum. As a result, the Euclidean geometry Pythagorean Theorem does not hold in spacetime. Spacetime is thus not Euclidean (Mazenc and Bernard, 2011). Flat-hyperbolic geometry is a term used to describe the new form of geometry represented by Eqn. (13). Instead of calling spacetime hyperbolic, we state it is non-Euclidean because hyperbolic geometry is the second term for the non-Euclidean geometry of Nikolai Ivanovich Lobachevski (1793–1856). Space is Euclidean by itself, while spacetime is not. As we will see soon, the reality that spacetime is not Euclidean explains the seemingly unusual properties of the contraction of length and the dilation of time. All disparities between time and space are based on the negative sign in Eqn. (13) (Ledrappier, 1981).

Due to the negative sign in Eqn. (13), $(\Delta s)^2$ might be either plus, minus, or zero. $(\Delta s)^2$ is positive when $(\Delta\tau)^2$ is greater than $(\Delta x)^2$. The world line in spacetime is termed as time-like and is seen in the future light cone since the time term predominates. When $(\Delta x)^2$ is greater than $(d\tau)^2$, $(\Delta s)^2$ is minus. The world line is termed as spacelike because the term space predominates in such a case. Figure 2.4 shows a spacelike universe exterior region of the light cone known as elsewhere. Here $(\Delta s)^2$ is zero when $(\Delta x)^2 = (\Delta\tau)^2$, in this example $(\Delta x) = \Delta\tau = (cdt)$. As a result, $\Delta x = cdt$, or $\Delta x/\Delta t = c$ (Tang et al., 2019).

However, velocity is $\Delta x/\Delta t$. It should be the world line of anything traveling at the light speed that is equal to *c*. Be a result, $(\Delta s)^2 = 0$ denotes a light ray, and the world line is referred to as light like. The light cone is made up of global lines that resemble light. Another feature of Euclidean space is that the shortest distance between two locations is a straight line. We will see now that the straight line is the largest distance among two places in non-Euclidean spacetime. Figure 2.6(a) shows the distance traveled using two space paths (Pelikan, 1984).

Using the Pythagorean Theorem, we can calculate the distance traveled along with pathway AB in Euclidean space.

$$S_{AB} = \sqrt{\left(\frac{y}{2}\right)^2 + x^2}$$

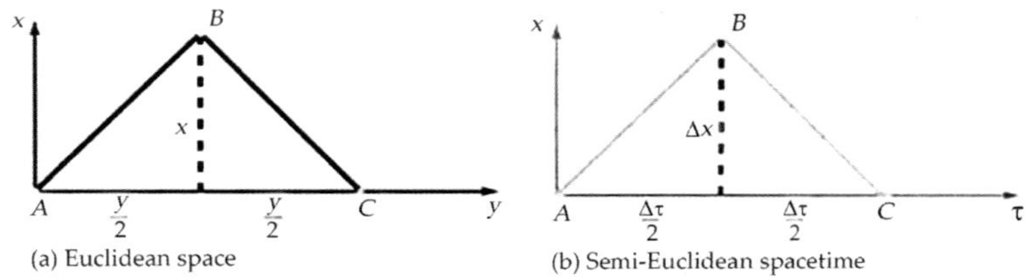

Figure 2.6. Space vs. spacetime.

Source: *https://sci.esa.int/web/lisa-pathfinder/-/56434-spacetime-curvature.*

Likewise, the distance along pathway BC is the same (Chen et al., 2006).

$$S_{Bc}=\sqrt{(\frac{y}{2})^2+x^2}$$

As a result, the cumulative distance covered along pathway ABC is:

$$S_{ABC}=S_{AB}+S_{Bc}=2\sqrt{(\frac{y}{2})^2+x^2} \quad (14)$$

The cumulative distance covered along pathway AC is:

$$S_{ABC}=\sqrt{y^2-(2x)^2}$$

But since,

$$S_{AC}=\frac{y}{2}+\frac{y}{2}=\text{y}$$

$$\sqrt{\pm^2-(2\ )^2}> \quad (15)$$

As predicted, the pathway ABC is longer as compared to the straight-line travel at the point of AC. Now consider the identical issue in spacetime, as seen in Figure 2.6(b). The subscript 0 is utilized on τ to denote that the clock is at rest. A clock's appropriate time is the time it reads while it is at rest. However, because this appropriate time is also equivalent to the spacetime interval, which is a constant, the interval computed along every time like world-line has been equivalent to its appropriate time (Chen et al., 2005). If a body is transported from point A to point B, the time elapsed on the clock as it goes from point A to point B is dsAB, and the time elapsed along pathway BC is dsBC. As a result of the calculation, the time spent on pathway ABC is less than the time spent on pathway AC. If two clocks began synchronizing at point A, they will read various times when they meet at the point of C (Rautaharju and Zhang, 2002). As a result, time, like distance, is often referred to as a route-dependent variable. An accelerated route is represented by ABC. (However, the acceleration at point B happens virtually instantly.) As a result, the appropriate time for an accelerating observer is shorter as compared to an observer at the position of rest. As a result, time should sluggish down while acceleration, as we shall see in our research of general relativity (Pitman, 2002).

A twin is an astronaut and journeyed to external space, whereas the other twin stayed on earth. The Lorentz dilation of time equation demonstrated that the returning astronaut will be younger than his twin who remained at home. Figure 2.6(b) is a spacetime representation of the twin paradox. The world line across spacetime for the twin who stays at home is represented by pathway AC, while the world line for the astronaut is represented by pathway ABC (Petras et al., 2004). Pathway ABC across spacetime is curved because, to return to Earth, the astronaut had an acceleration phase. The astronaut may thus no longer be regarded as an inertial observer. Because the pathway AC of the twin who stays at home is a straight line in spacetime, she has been an observer with inertial motion. The time elapsed alongside pathway ABC, the astronaut's journey is shorter than the time elapsed along with pathway AC, the stay-at-home's passageway as demonstrated in the previous paragraph. Consequently, the astronaut returns home younger than his twin who remained at home (Kolev and Petrakieva, 2005).

Among the most essential properties of the constant interval is that it permits us to create a decent geometrical representation of spacetime as experienced by various individuals. Figure 2.7 illustrates a piece of spacetime for a static observer S. The orthogonal axes represent x and τ coordinates. The light lines OL' and OL are angled at 45°. The interval, Eqn. (13), is generated for a succession of x and τ values and appears in the illustration as the family of hyperbolas (Chen et al., 2013). (The intervals will have been a series of concentric circles around the origin of O rather than such hyperbolas if spacetime were Euclidean.) The light cone future is defined by hyperbolas drawn around the -axis, whereas hyperbolas drawn around the x-axis are defined elsewhere. Within the light cone, the interval has negative values and positive values. A point of reference S, traveling at the velocity v, will have the worldline of its origin (straight line) across spacetime inclination angle θ given by (Bravo et al., 2005):

$$\theta = \tan^{-1}\frac{v}{c}$$

For instance, if S is traveling at c/2, then θ = 26.6°. Figure 2.7 depicts this world line. However, the world line of coordinates (x' = 0) is identified as the time axis τ' of the S' frame in the diagram. When τ' coincides with the hyperbolas family at Δs = 1, 2, 3, …, the time scale alongside the τ'-axis becomes τ' = 1, 2, 3, …, (Remember that $(\Delta s)^2 = (\Delta\tau')^2 - (\Delta x')^2$, and the coordinate system's origin is Δx' = 0, therefore Δs = Δτ') It is worth noting that the scale on the τ'-axis differs from the scale on the τ-axis (Hooper, 2015).

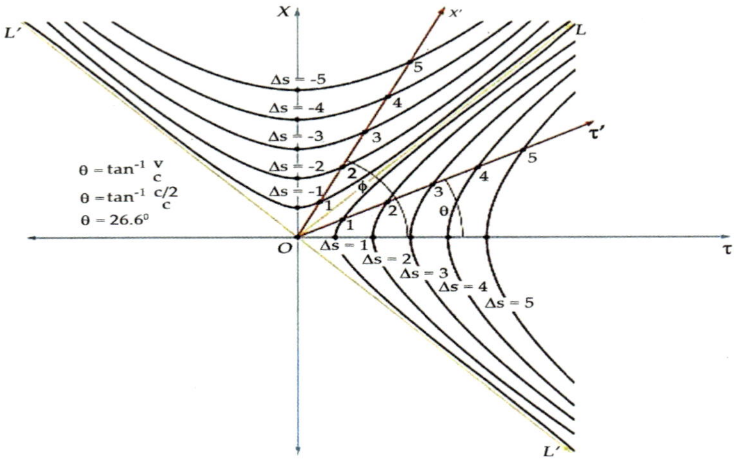

Figure 2.7. The constant interval on a spacetime map.

Source: https://logosconcarne.com/2019/03/18/sr-x3-spacetime-interval/.

To construct the x'-axis on this graph, keep in mind that it depicts all points when τ' = 0 (Figure 2.8) (Le Mézo et al., 2017).

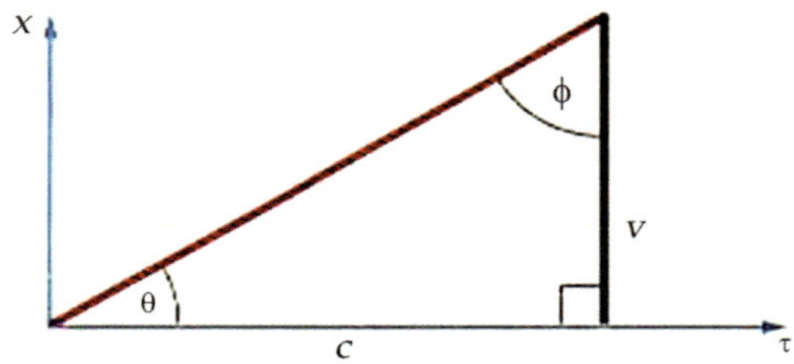

Figure 2.8. Determining the slope of the x'-axis.

Source: https://www.onlinemath4all.com/slope-of-a-line-parallel-to-x-axis.html.

The angle is determined using the equation. This S' frame is exclusive for a certain value of v. The coordinate system of another inertial observer traveling at varying speeds will also be distorted. Consequently, the angle θ will vary based on the value of *v* (Tahavori and Shaker, 2013).

The movement of the inertial observer S” appears to distort the basic orthogonal spacetime into a spacetime that is skewed. This skewed spacetime provides a straightforward explanation for the contraction of length and dilation of time. Figures 2.9 are a collection of spacetime illustrations dependent upon the constant interval, illustrating the contraction of length, dilation of time, and simultaneity (Zweimüller, 1998).

Figure 2.9 depicts a 4.00-unit-long rod at rest in a rocket ship S’ traveling at speed c/2. The top of the stick in S’ has a world line that is parallel to the τ’-axis. (In S,’ each line parallel to the τ’-axis has just one value of x’ and hence depicts an object at rest) (Bruin et al., 2008).

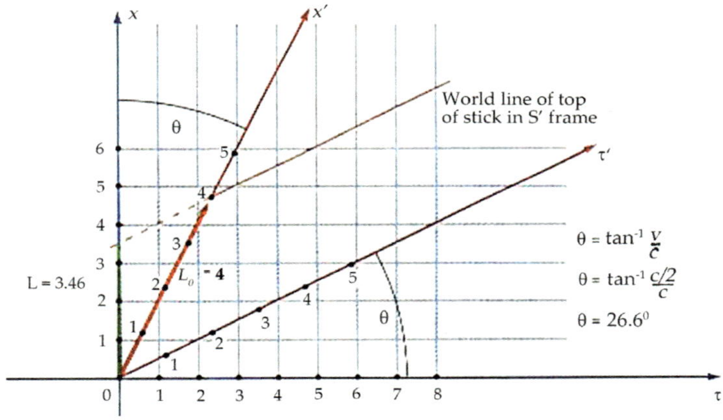

Figure 2.9. The contraction of length, rod at rest in S’ frame.

Source: http://hyperphysics.phy-astr.gsu.edu/hbase/Relativ/tdil.html.

The world line meets the x-axis at x= 3.46, that is the length of the rod L, as perceived through the S-frame observer when dashed back to the x-axis. To the spectator on Earth, the S frame, and the rod at the position of rest in the traveling rocket frame seems constricted. The Lorentz contraction is, of course, the contraction of the traveling rod. It is easy to see with the spacetime diagram (Xu et al., 2020).

The identical Lorentz contraction is seen in Figure 2.10, but this time from the S’ frame. In the S frame, the earth, a rod 4.00 unit’s length L_0 is at rest (Ruelle, 2005).

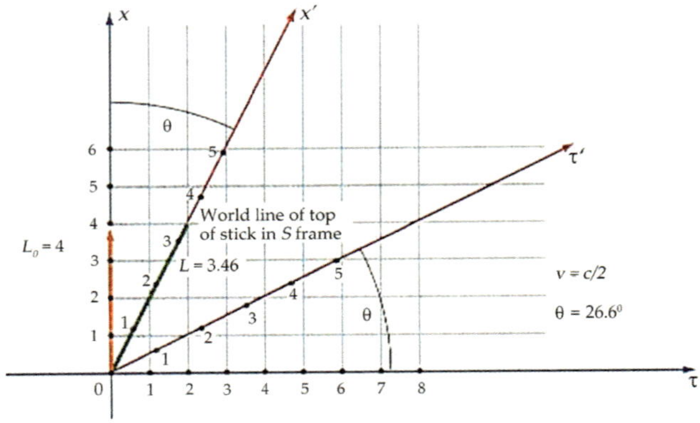

Figure 2.10. The contraction of length, rod at rest in S frame.

Source: http://electron6.phys.utk.edu/PhysicsProblems/Mechanics/8-Relativity/properlength.html.

An observer in the rocket ship frame, also known as the S' frame, considers himself to be at rest while the earth is moving away from him at a velocity of –v. As the world line, that radiates from the top of the rod, coincides with the astronaut's coordinate system, the astronaut observes it (Dubins and Freedman, 1966). Using the S frame, as indicated in the diagram, we may determine how long the rod is that he perceives by drawing the world line of the top of the rod. The intersection of such a world line with the x'-axis occurs at the coordinates X' = 3.46. As a result, the rocket observer determines that the rod on Earth is just 3.46 units long, which is the length L. As a result, the observer from the rocket ship notices a similar length contraction.

The non-Euclidity of spacetime is the underlying cause of such contractions (Góra, 2009). With the help of the spacetime diagram, shown in Figure 2.10, we can clearly understand the impact of time dilation. It is the location x' = 2 that a clock is at the position of rest in a traveling rocket ship. Its worldline is drawn parallel to the τ'-axis, as seen in the diagram (Romig et al., 2019). According to the figure, time elapses among the occurrences of events A and B on the S" clock, which is equal to dτ' = 4.0 – 2.0 = 2.0. By moving the dashed lines from event, A and event B downward to the τ-axis, we may find out how long this time interval lasts as perceived by the S frame of the earthman. (While these lines are parallel to the x-axis, they are also perpendicular to the y-axis since S is an orthogonal frame) (Habib et al., 1991).

REFERENCES

1. Ahn, H. S., Chen, Y., & Podlubny, I., (2007). Robust stability test of a class of linear time-invariant interval fractional-order system using Lyapunov inequality. *Applied Mathematics and Computation, 187*(1), 27–34.
2. Ahsan, Z., & Ali, M., (2017). Curvature tensor for the spacetime of general relativity. *International Journal of Geometric Methods in Modern Physics, 14*(05), 1750078.
3. Ahsan, Z., (2005). The spacetime of general relativity. *Bull. Cal. Math. Soc., 97*(3), 191–200.
4. Almeida, R., & Ledberg, A., (2010). A biologically plausible model of time-scale invariant interval timing. *Journal of Computational Neuroscience, 28*(1), 155–175.
5. Bowen, R., (1979). Invariant measures for Markov maps of the interval. *Communications in Mathematical Physics, 69*(1), 1–14.
6. Bravo, J. M., Limón, D., Alamo, T., & Camacho, E. F., (2005). On the computation of invariant sets for constrained nonlinear systems: An interval arithmetic approach. *Automatica, 41*(9), 1583–1589.
7. Brill, D., & Jacobson, T., (2006). Spacetime and Euclidean geometry. *General Relativity and Gravitation, 38*(4), 643–651.
8. Bruin, H., Rivera-Letelier, J., Shen, W., & Van, S. S., (2008). Large derivatives, backward contraction and invariant densities for interval maps. *Inventiones Mathematicae, 172*(3), 509–533.
9. Chambon, E., Burlion, L., & Apkarian, P., (2016). Overview of linear time-invariant interval observer design: Towards a non-smooth optimization-based approach. *IET Control Theory & Applications, 10*(11), 1258–1268.
10. Chen, S. H., & Chou, J. H., (2013). Robust controllability of linear time-invariant interval systems. *Journal of the Chinese Institute of Engineers, 36*(5), 672–676.
11. Chen, Y., Ahn, H. S., & Podlubny, I., (2005). Robust stability check of fractional order linear time-invariant systems with interval uncertainties. In: *IEEE International Conference Mechatronics and Automation* (Vol. 1, pp. 210–215). IEEE.
12. Chen, Y., Ahn, H. S., & Xue, D., (2006). Robust controllability of interval fractional order linear time-invariant systems. *Signal Processing, 86*(10), 2794–2802.

13. Crutchfield, J. P., & Mitchell, M., (1995). The evolution of emergent computation. *Proceedings of the National Academy of Sciences, 92*(23), 10742–10746.
14. Deshko, Y., (2022). Spacetime. In: *Special Relativity* (Vol. 1, pp. 45–72). Springer, Cham.
15. Dikshit, B., (2013). Space-time diagram approach in derivation of Lienard–Wiechert potential for a moving point charge. *Canadian Journal of Physics, 91*(7), 519–521.
16. Dray, T., (2013). Using three-dimensional spacetime diagrams in special relativity. *American Journal of Physics, 81*(8), 593–596.
17. Dubins, L. E., & Freedman, D. A., (1966). Invariant probabilities for certain Markov processes. *The Annals of Mathematical Statistics, 37*(4), 837–848.
18. During, E., (2012). On the intrinsically ambiguous nature of spacetime diagrams. *Spontaneous Generations: A Journal for the History and Philosophy of Science, 6*(1), 160–171.
19. Ellis, G. F., (2006). Physics in the real universe: Time and spacetime. *General Relativity and Gravitation, 38*(12), 1797–1824.
20. Favaron, E., Ince, C., Hilty, M. P., Ergin, B., Van, D. Z. P., Uz, Z., & Endeman, H., (2021). Capillary leukocytes, microaggregates, and the response to hypoxemia in the microcirculation of coronavirus disease 2019 patients. *Critical Care Medicine, 49*(4), 661.
21. Flores, F. J., (2007). Communicating with accelerated observers in Minkowski spacetime. *European Journal of Physics, 29*(1), 73.
22. Geroch, R., (1968). What is a singularity in general relativity?. *Annals of Physics, 48*(3), 526–540.
23. Góra, P., (2009). Invariant densities for piecewise linear maps of the unit interval. *Ergodic Theory and Dynamical Systems, 29*(5), 1549–1583.
24. Habib, M., Kelly, D., & Möhring, R. H., (1991). Interval dimension is a comparability invariant. *Discrete Mathematics, 88*(2, 3), 211–229.
25. Hameroff, S., (2001). Consciousness, the brain, and spacetime geometry. *Annals of the New York Academy of Sciences, 929*(1), 74–104.
26. Hooper, W. P., (2015). The invariant measures of some infinite interval exchange maps. *Geometry & Topology, 19*(4), 1895–2038.

27. Hu, S., & Wang, J., (2000). On stabilization of a new class of linear time-invariant interval systems via constant state feedback control. *IEEE Transactions on Automatic Control, 45*(11), 2106–2111.
28. Janis, A. I., (2018). On mass, spacetime curvature, and gravity. *The Physics Teacher, 56*(1), 12, 13.
29. Kolev, L., & Petrakieva, S., (2005). Assessing the stability of linear time-invariant continuous interval dynamic systems. *IEEE Transactions on Automatic Control, 50*(3), 393–397.
30. Le Mézo, T., Jaulin, L., & Zerr, B., (2017). An interval approach to compute invariant sets. *IEEE Transactions on Automatic Control, 62*(8), 4236–4242.
31. Ledrappier, F., (1981). Some properties of absolutely continuous invariant measures on an interval. *Ergodic Theory and Dynamical Systems, 1*(1), 77–93.
32. Leonardi, A. M., Mobilio, S., & Fazio, C., (2022). A teaching proposal for the didactics of special relativity: The spacetime globe. *Physics Education, 57*(4), 045002.
33. Leubner, C., (1981). Spacetime diagrams with base vectors. *American Journal of Physics, 49*(4), 368–371.
34. Mazenc, F., & Bernard, O., (2011). Interval observers for linear time-invariant systems with disturbances. *Automatica, 47*(1), 140–147.
35. Pelikan, S., (1984). Invariant densities for random maps of the interval. *Transactions of the American Mathematical Society, 281*(2), 813–825.
36. Petras, I., Chen, Y., Vinagre, B. M., & Podlubny, I., (2004). Stability of linear time-invariant systems with interval fractional orders and interval coefficients. In: *Second IEEE International Conference on Computational Cybernetics, 2004: ICCC 2004* (Vol. 1, pp. 341–346). IEEE.
37. Pitman, J., (2002). Poisson–Dirichlet and GEM invariant distributions for split-and-merge transformations of an interval partition. *Combinatorics, Probability and Computing, 11*(5), 501–514.
38. Puig, V., Saludes, J., & Quevedo, J., (2003). Worst-case simulation of discrete linear time-invariant interval dynamic systems. *Reliable Computing, 9*(4), 251–290.
39. Rautaharju, P. M., & ZHANG, Z. M., (2002). Linearly scaled, rate-invariant normal limits for QT interval: Eight decades of

incorrect application of power functions. *Journal of Cardiovascular Electrophysiology, 13*(12), 1211–1218.

40. Romig, S., Jaulin, L., & Rauh, A., (2019). Using interval analysis to compute the invariant set of a nonlinear closed-loop control system. *Algorithms, 12*(12), 262.
41. Rousseaux, G., Mathis, C., Maïssa, P., Philbin, T. G., & Leonhardt, U., (2008). Observation of negative-frequency waves in a water tank: A classical analogue to the Hawking effect?. *New Journal of Physics, 10*(5), 053015.
42. Ruelle, D., (2005). Differentiating the absolutely continuous invariant measure of an interval map f with respect to f. *Communications in Mathematical Physics, 258*(2), 445–453.
43. Tahavori, M., & Shaker, H. R., (2013). Model reduction via time-interval balanced stochastic truncation for linear time-invariant systems. *International Journal of Systems Science, 44*(3), 493–501.
44. Tang, W., Wang, Z., Wang, Y., Raïssi, T., & Shen, Y., (2019). Interval estimation methods for discrete-time linear time-invariant systems. *IEEE Transactions on Automatic Control, 64*(11), 4717–4724.
45. Valand, R. S., (1968). Invariant interval estimation of a location parameter. *The Annals of Mathematical Statistics, 39*(1), 193–199.
46. Wei, K., (1994). Stabilization of linear time-invariant interval systems via constant state feedback control. *IEEE Transactions on Automatic Control, 39*(1), 22–32.
47. Witten, E., (2001). Reflections on the fate of spacetime. *Physics Meets Philosophy at the Planck Scale, 1*(2), 125–137.
48. Xu, R., He, F., & Wang, B. Y., (2020). Interval counterexamples for loop invariant learning. In: *Proceedings of the 28th ACM Joint Meeting on European Software Engineering Conference and Symposium on the Foundations of Software Engineering* (Vol. 1, pp. 111–122).
49. Zweimüller, R., (1998). Ergodic structure and invariant densities of non-Markovian interval maps with indifferent fixed points. *Nonlinearity, 11*(5), 1263.

CHAPTER 3

QUANTUM PHYSICS

CONTENTS

3.1. INTRODUCTION

We viewed particles as small, brittle spheres of substance, whereas a wave was a disruption propagated by a medium. Figure 3.1 illustrates one of the most notable distinctions between the two conceptions. In Figure 3.1(a), two particles hit each other and bounce off one another, and subsequently travel in a different direction (Piron, 1976; Busch et al., 1997).

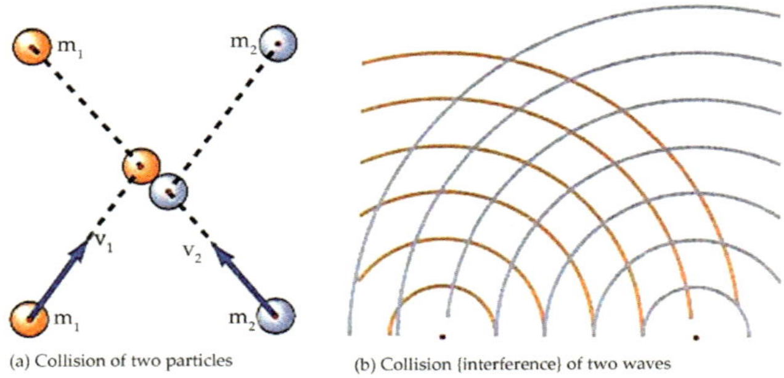

Figure 3.1. The properties of waves and particles.

Source: http://hyperphysics.phy-astr.gsu.edu/hbase/mod1.html.

Both waves interact in Figure 3.1(b), and they do not bounce off one another. They combine due to the concept of superposition, and then every wave is restored to its normal path as though the waves had never met (Simon, 1979; Zeilinger, 1999).

Another distinction between a wave and a particle is that the particle's complete energy is "concentrated" in its confined mass. The energy of a wave, on either side, is distributed over the whole wave. As a result, there is a considerable distinction between a wave and a particle.

Light is an electromagnetic wave, as we have seen. Polarization, diffraction, and interference are all common wave phenomena that are investigated and validated several times in the laboratory (Popper, 1950; Arndt et al., 2009). However, certain perceived conflicts with the waveform of light have emerged throughout time. The following three scientific effects would be discussed:

- Compton scattering;
- Radiation from blackbodies;
- The photoelectric effect.

3.2. BLACKBODY RADIATION

Radiation is emitted and absorbed by all things. (For instance, electromagnetic waves are used to transport heat.) The law of Stefan-Boltzmann states that the quantity of energy emitted is directly proportional to the 4th power of a temperature, but it did not specify how the heat emitted was a function of the wavelength. We will anticipate the energy to be dispersed evenly between all potential wavelengths since the radiation is made up of electromagnetic waves. The distribution of energy, meanwhile, is not uniform and changes with frequency and wavelength (Giamarchi, 2003; Greenberger et al., 2009).

All traditional methods to compensate for the distribution of energy failed. Consider how a body may emit energy for a minute. An electromagnetic wave is produced by an alternating electric charge, as we all know. As seen in Figure 3.2, a body is made up of a huge number of atoms arranged in a lattice pattern (a). The positively ionized atom is placed at the lattice structure in a metallic material, and the atom's outer electron flows around the lattice as a component of the electron gas (Barenco, 1996; Economou, 2006).

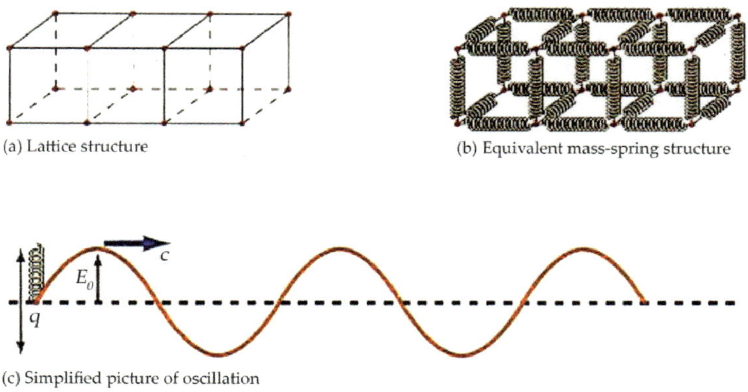

Figure 3.2. Electromagnetic radiation is emitted by a solid material.

Source: https://courses.lumenlearning.com/physics/chapter/14-7-radiation/.

Under the influence of all the forces from all its adjacent atoms, every atom of the lattice structure is in a condition of stability. The atom is free to oscillate about this point of balance. Figure 3.2(b) depicts a mechanical equivalent of the structure as a sequence of masses linked through springs. Every mass may fluctuate around its stability position. Further to clarify the concept, consider one ionized atom having a charge q oscillating in simple harmonic motion (SHM), as illustrated in Figure 3.2(c). (Gel'fand and Yaglom, 1960; Walmsley and Rabitz, 2003). The body emits an electromagnetic wave generated by the fluctuating charge. Every single ionized atom oscillates, with its set frequency and corresponding radiation. The body produces radiation of all such various frequencies since it is composed of millions of such fluctuating charges, thus the emission of the spectrum must be continuous. The amplitude of the oscillation finds out the intensity of the radiation. As you may remember from elementary physics, the equation for a common emitted wave is:

$$E = E_0 \sin (kx - \omega t)$$

Whereas,

$$k = 2\pi/\lambda$$

and,

$$\omega = 2\pi v$$

The electromagnetic wave's frequency is hence the frequency of the fluctuating charge. The amplitude of wave E_0 is determined by the amplitude of the alternating charge's SHM. The thermal energy causes the ionized atoms to oscillate having increased amplitude around their equilibrium state whenever the body is warmed (Holevo, 1973; Eisberg and Resnick, 1985). The energy density of the waves that are released may be calculated as:

$$u = \varepsilon_0 E_2$$

or,

$$u = \varepsilon_0 E_0^2 \sin^2(k_x - \omega_t) \qquad (1)$$

As a result, as the amplitude of the oscillation E_0 grows, extra energy is released. When a heated body is left alone, it releases energy into the environment through the procedure of radiation, and the oscillation's amplitude drops. The energy of an electromagnetic wave is determined by the amplitude of the oscillation. All amplitude of oscillation of the lattice is feasible, and therefore all potential frequencies are present, due to the unusually huge quantity of ionized atoms in the crystalline lattice that may contribute to the vibrations. As a result, the classical model of blackbody radiation allows for all frequencies and energy for electromagnetic waves. Furthermore, the experiment contradicts this traditional understanding (Wilcox, 1967; Gazeau, 2009).

A German scientist named Max Planck (1858–1947) attempted to "fit" the experimental evidence to the theory. Although, he saw that he needed to defy convention and suggest a novel and innovative notion. Planck thought that the oscillators of the atom might not take on all potential energies but might only oscillate through discrete quantities of energy determined by Planck's equations.

$$E = nh\nu \qquad (2)$$

wherein; "h" is an invariant, currently known as *Planck's constant*, with a value of:

$h = 6.625 \times 10^{-34}$ Js

In Eqn. (2), v is the oscillator's frequency and n is an integer, a quantity now known as a quantum number. It is presently believed that the energy of vibrating atoms is quantized, or confined to the values provided by Eqn. (2). Therefore, the atom may have energies like hv, $2h\nu$, $3h\nu$, etc., but never 2.5 $h\nu$. Quantization is fundamentally incompatible with traditional electromagnetic theory. Under the classical theory, as the oscillating charge emits energy, it releases energy, and the amplitude of the oscillation gradually diminishes (Guhr et al., 1998; Deutsch et al., 2000). If the oscillator's energy is quantified the amplitude may not reduce continuously, and the vibrating charge may not emit whereas in this quantum condition. If the oscillator now loses one quantum state of energy, the energy gap between the two states can be emitted away. The supposition of discrete energy levels implies that the radiation procedure may happen when the oscillator transitions from the first quantized energy state to another quantized energy state (Geiger, 2018; French and Redhead, 1988).

For instance, if the oscillating charge is in quantum state 4, it possesses energy.

$$E_4 = 4h\nu$$

The oscillator has the energy whenever it falls to quantum state 3.

$$E_3 = 3h$$

To emit energy, the oscillator must descend from its fourth state to its third state.

$$\Delta E = E_4 - E_3 = 4h\nu - 3h\nu = h\nu$$

As a result, the quantity of energy emitted is constantly in extremely tiny packets of energy. A quantum of energy was the name given to this little bundle of emitted electromagnetic energy. A photon was eventually given to this bundle of electromagnetic energy (Stapp, 1999; Ritz et al., 2002).

Even though this quantum hypothesis led to the right description of black body radiation, it left many concerns unsolved. Why must the oscillator's energy be quantized? How does the energy from the black body become spread out into Maxwell's electromagnetic wave if it is released as a small packet of energy? How does the energy that is scattered out in the wave become compacted back into the small quantum of energy that an atomic oscillator may absorb? These and other concerns bothered Planck and the rest of the physics community. Even though Planck founded quantum mechanics and was awarded the Nobel Prize for it, he spent several years attempting to refute his basic theory (Morandi, 1992; Michelini et al., 1992).

3.3. THE PHOTOELECTRIC EFFECT

In 1887, while conducting experiments to confirm the presence of electromagnetic waves, Heinrich Hertz discovered that when light is incident on a substrate of metal, the surface released electrical charges, measured in the form of electrical currents. This phenomenon is known as the photoelectric effect. The photoelectric effect was the first evidence that light is composed of photons. Philip Lenard's further investigations in 1900 showed that such electrical charges where electrons, sometimes called photoelectrons. A schematic representation seen in Figure 3.3, best exemplifies the photoelectric effect. By flipping the switch S, the anode of the phototube becomes +ve and the cathode becomes –ve. Monochromatic light having intensity I_1 is permitted to shine on the photo tube's cathode, allowing electrons to be released (Brout et al., 1995; Schwartz et al., 2005).

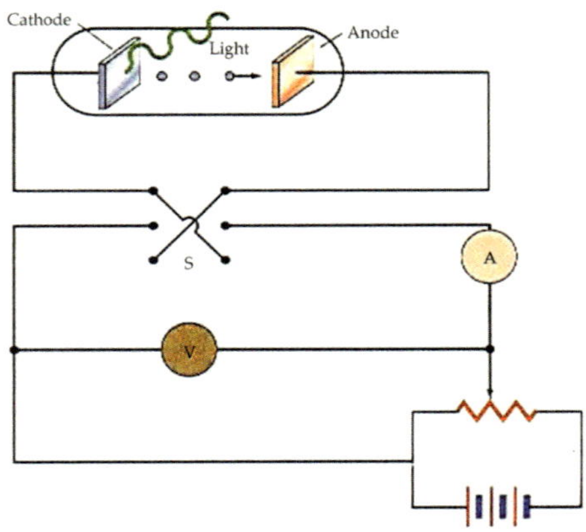

Figure 3.3. Diagrammatic representation of the photoelectric effect.

Source: https://www.researchgate.net/figure/Photoelectric-Effect-Schematic_fig1_258514434.

Such electrons are attracted to the anode, where they subsequently travel across the connected circuit. The circuit's ammeter monitors such current. Beginning with a +ve potential *V*, the current is measured as the *V* decreases. When the voltage *V* is decreased to zero, the switch *S* is flipped such that the anode becomes +ve and the cathode becomes –ve. As photo electrons approach the anode, the –ve anode now repelled them. If such potential is increased evermore –ve, the kinetic energy (KE) of the electrons will finally be insufficient to achieve the –ve stopping potential, and no additional electrons will approach the anode. Consequently, the current value of "*i*" becomes 0. Figure 3.4 depicts the current "*i*" in the circuit as a function of the voltage among the plates (Gallagher and Cooke, 1979; French, 1989).

If we raise the light intensity to I_2 and repeat the experiment, the second curve seen in the picture is obtained.

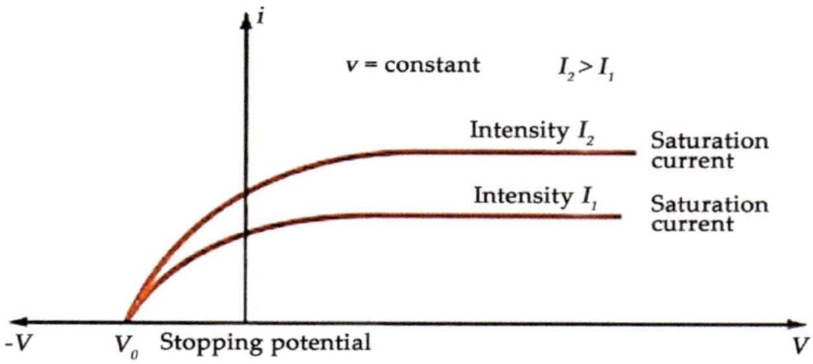

Figure 3.4. The relationship between current "i" and voltage V for the photoelectric effect.

Source: https://byjus.com/jee/photoelectric-effect/.

Whenever the value of *V* is positive and higher, Figure 3.4 shows that the current "*i*" is invariant. That happens because all photoelectrons created at the cathode reach the anode. We achieve a greater invariant value of current by raising the intensity I, as more photoelectrons have been released per unit of time. This demonstrates that the released quantity of electrons (the current) is directly proportional to the intensity of the incoming light (Boyer, 1969; Smoot et al., 1977).

$i \propto I$

Remember that whenever the potential is decreased to 0, the tube still contains a current. Although there has been no electric field to attract them to the anode, a significant number of photoelectrons approach the anode due to the KE they contain whenever they exit the cathode. Even as the switch S in Figure 3.3 is flipped, the potential V between the plates becomes –ve and photoelectrons tend to be eliminated. As the hindering potential V is composed of a –ve charge, the current "i" declines (Figure 3.4), showing that relatively fewer photoelectrons have been approaching the anode. When V is dropped to V_0, zero current flows through the circuit; V_0 is referred to as the stopping potential (Ford et al., 1985). Remember that the value remains constant regardless of the intensity. (At V_0, both curves overlap.) Therefore, the stopping potential is irrespective of light intensity or expressed in another manner, which has not been a function of the intensity of light. This is expressed as,

$$V_0 \neq V_0(I) \qquad (3)$$

The photoelectrons' KE is connected to the hampering potential. The electron's KE should be equivalent to the potential energy (PE) in between plates for this to approach the anode. (An analogy to mechanics could be useful at this time.) If we want to toss a ball up to a height of h, and that would have the PE, we may use the formula:

$$\text{PE} = mgh$$

The starting velocity of the ball should be v_0 so that the starting KE of the ball has been the same as its ultimate PE.

$$\tfrac{1}{2}\,\text{KE} = mv_o$$

Therefore, the electron's KE should equal the PE in between plates, or,

$$\text{KE} = eV \qquad (4)$$

wherein; e represents the electron's charge and V represents the voltage in between plates. The hampering potential affects electrons whose KE is smaller than that calculated using Eqn. (4). Even the most excited electrons (those that have the highest KE) cannot approach the anode whenever $V = V_0$, the stopping potential (Dicke et al., 1965; Boyer, 2018).

Thus,

$$\text{KE}_{\text{max}} = eV_0 \qquad (5)$$

The maximal KE of photoelectrons is not a function of the intensity of incident light, as shown by Eqns. (3) and (5):

$$\text{KE}_{\text{max}} \neq \text{KE}_{\text{max}}(I)$$

Experiments have also revealed that there has been virtually no time lag between when light appears on the cathode and when photoelectrons are released.

The curves displayed in Figure 3.5 are obtained by keeping the intensity invariant and performing the experiment using various light frequencies (Cooke and Gallagher, 1980; Chang and Rhee, 1984). The saturation current (maximal current) is similar to every light frequency as well as the intensity is invariant, as shown in Figure 3.5.

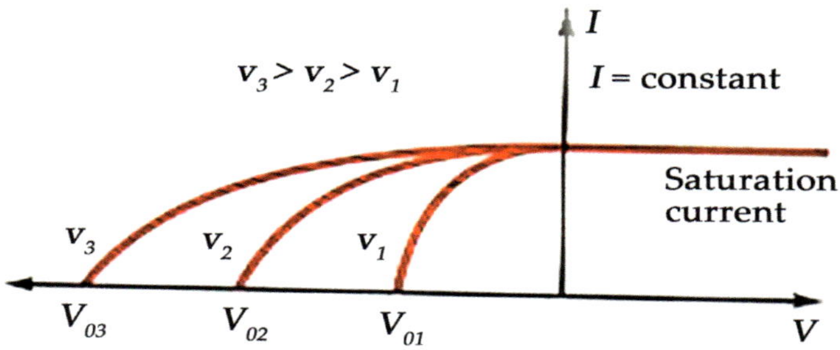

Figure 3.5. For various frequencies of light, current "i" as a function of voltage V.

Source: https://www.snapsolve.com/solutions/Thegraphs-show-the-variation-of-current-I-y-axis-in-two-pho-tocell-A-and-B-as-a--1681404329954305.

The stopping potential, on the other hand, varies depending on the frequency of the incoming light. Because, according to Eqn. (5), the stopping potential is directly proportional to the maximal KE of the photoelectrons, the maximal KE of the photoelectrons must be directly proportional to the frequency of the incoming light. Photoelectrons have the highest possible KE, which is displayed as a function of frequency in the graph depicted in Figure 3.6. As a preliminary observation, it should be noted that the maximal KE of the photoelectrons is directly proportional to the frequency of the input light source (Cornelius and Dowling, 1999; Reiser and Schächter, 2013).

i.e., $KE_{max} \propto \mathbf{v}$

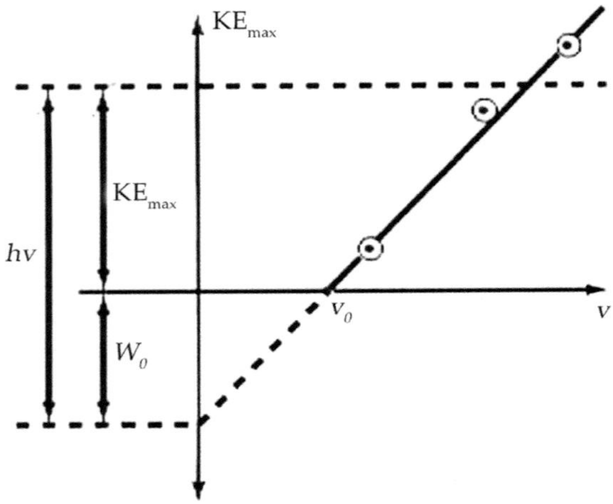

Figure 3.6. Maximum kinetic energy (KE_{max}) as a function of frequency ν.

Source: https://courses.lumenlearning.com/physics/chapter/29-2-the-photo-electric-effect/.

The second observation is that there is a threshold frequency v_0, below which photoelectronic discharge does not occur. Thus, no photoelectric effect happens until the frequency of the incoming light is greater than the cutoff frequency v_0. The v_0 resides in the UV area of the spectrum for the majority of metals, although in the visible range for alkali metals (Greenstein and Hartke, 1983; Reiser and Schächter, 2013).

3.3.1. The Photoelectric Effect Cannot Be Explained by Classical Electromagnetism

The photoelectric effect had been first attempted to be explained using the classical theory of electromagnetic. Table 3.1 compares the experiment's outcomes to the expectations of classical electromagnetic theory. The assumption that the photo-current is directly proportional to the intensity of the incoming light is the lone point where the experiment and theory coincide (Dunbar and McMahon, 1998; Haslinger et al., 2018).

Table 3.1. The Photoelectric Effect

Experimental Outcomes	Assumptions of Classical Electromagnetism Based on Theory	Agreement
$KE_{max} \neq KE_{max}(I)$	$KE_{max} \propto I$	No
Cutoff frequency ν0	There must not be a cutoff frequency	No
$i \propto I$	$i \propto I$	Yes
$KE_{max} \propto \nu$	KE_{max} not $\propto \nu$	No
No time lag for emanation of electrons	There must be a time lag	No

There must be no minimum threshold frequency v_0 for photoelectron discharge, as per classical theory. The experimental data contradict this assumption (Dunbar and McMahon, 1998; Middelmann et al., 2012).

Energy is spread evenly over the whole electric wavefront, as per the theory of classical electromagnetic. Whenever the wave meets the electron on the cathode, the electron must only take a small percentage of the overall wave's energy. As a result, there must be a time delay to allow the electron to collect sufficient energy before being expelled. It has been discovered in experiments that emanation happens instantly after lighting; there has been no temporal lag in emission. Ultimately, according to the theory of classical electromagnetic, highly intense light with rather lower frequencies would emit more than higher-frequencies light with very lower intensities. Once again, the theory contradicts the experimental outcomes. As a result, the photoelectric effect may not be explained by the theory of classical electromagnetic (Kelly, 1981; Safronova et al., 2013).

3.3.2. Photoelectric Effect Theory of Einstein

Einstein presented a novel and innovative answer to the issue of the photoelectric effect the same year he published his special theory of relativity, in 1905. Utilizing Planck's notion of energy quantization to solve the black bodies' radiation issue, Einstein thought that the energy of the electromagnetic waves had been condensed into Planck's small quanta of energy rather than being radiated evenly alongside the wavefront. Planck thought that the radiators of the atom had been quantized, but he still thought that when the wave travelled, the energy radiated over the wave. In contrast, Einstein hypothesized that as the wave advanced; the energy remained in the small quanta of energy that will eventually be recognized as the photon. The energy was therefore contained in the photon.

$E = h\nu$ (6)

Einstein hypothesized that an electron on the surface of the metal was impacted by such a focused bundle of radiant energy. This full quantum of energy ($E = h\nu$) was subsequently absorbed by the electron. The electron uses some of this energy to split away from the material, while the remainder becomes the electron's KE (Widger et al., 1976; Spencer et al., 1982), i.e.,

(The absorbed energy that is incident) – (energy to split away from solid) = (maximal Kinetic Energy of the electron) (7)

We refer to the energy required for an electron to escape a solid as the work function of the solid and represent it by the symbol W_0. We may formally formulate Eqn. (7) like:

$E - W_0 = KE_{max}$(8)

or,

$h\nu - W_0 = KE_{max}$ (9)

The final maximal KE of photoelectrons is found using Eqn. (9) as:

$KE_{max} = h\nu - W_0$ (10)

Einstein's photoelectric Eqn. (10) is well-known (Gerlach, 1976; Safronova et al., 2011). Figure 3.6 shows that when the photoelectrons' KE_{max} is equivalent to 0, the frequency ν equals the threshold frequency ν_0. As a result, Eqn. (10) becomes:

$0 = h\nu_0 - W_0$

As a result, we may express the metal's work function as:

$W_0 = h\nu_0$ (11)

As a result, the photoelectric equation of Einstein may alternatively be written as:

$KE_{max} = h\nu - h\nu_0$ (12)

There has been no photoelectric effect at light frequencies equivalent to or even less than ν_0 since the incoming wave does not have enough energy to release the electron from the solid. This illustrates why the photoelectric effect disappears below a certain frequency (Sorokin et al., 2007; Gerlach, 1976).

There was not sufficient quantitative evidence available to prove Einstein's theory of the photoelectric effect whenever he suggested it. R. A. Millikan validated Einstein's hypothesis of the photoelectric effect in 1914 with tests (like the experiment detailed here).

The vanishing of a time lag for photoelectronic emanation is explained by Einstein's theory. When a photon strikes an electron on the surface of the metal, the electron acquires sufficient energy to release it instantaneously (Adawi, 1964; Klassen, 2011). Einstein's equation also closely approximates that the photoelectron's maximal KE is directly proportional to the frequency of incoming light. As a result, Einstein's equation closely approximates experimental outcomes. Their photoelectric effect theory of Einstein is noteworthy since it had been the first utilization of quantum principles. The light must be thought of as having both a particle and a wave nature. (A photon is a particle of light.) Einstein was awarded the Nobel Prize in Physics in 1921 for his interpretation of the photoelectric phenomenon. As previously stated, Einstein's study on the photoelectric effect had been released around a similar time as his special relativity paper in 1905 (Pratt et al., 1973; McKagan et al., 2007).

As a result, he had been considering both concepts at the same moment. It is no surprise, therefore, that he had been unconcerned with the ether's role in the transmission of electromagnetic waves. He did not require a medium for such waves to spread since he might now visualize light like a particle, a photon (Figure 3.7) (Lamb and Scully, 1968).

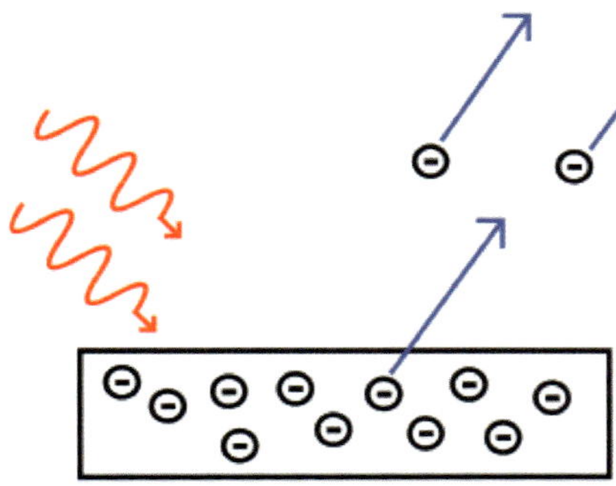

Figure 3.7. Light waves (red wavy lines) striking the surface of the metal cause electrons to be emitted from the metal in the photoelectric effect.

Source: https://www.khanacademy.org/science/physics/quantum-physics/photons/a/photoelectric-effect.

3.4. THE PROPERTIES OF THE PHOTON

The light should be a wave, as per classical physics. The photoelectric effect, on the other hand, requires light to be a particle, a photon. So, what exactly is light? Is light a particle or is light just a wave? If the light is a particle, this should contain a few of the properties of particles, such as energy, momentum, and mass (Steinberg et al., 1996; Ossiander et al., 2018). Considering the photon's mass for a moment. The relativistic particle mass was calculated using the equation:

$$m = \frac{m_o}{\sqrt{1 - \frac{v^2}{c^c}}}$$

However, as a light particle, the photon should travel at the speed of light, c.

As a result, its mass is given as:

$$m = \frac{m_o}{\sqrt{1 - \frac{v^2}{c^c}}} = m_o / 0$$

However, division by 0 is unknown. The only method to fix this difficulty is to consider a photon's idle mass to be 0.

Photon $m_0 = 0$ (13)

It may appear contradictory initially, but while the photon is constantly moving at the speed of light, this is never at rest and so does not require a rest mass. Eqn. (13) takes an indefinite form when m_0 equals 0 (Clauser, 1974; Agostini and Petite, 1988). Even though the mass of the photon may not be described by the equation, it may be derived from Eqn. (12):

$$E = mc^2$$

Consequently,

$$m = \frac{E}{C^2}$$

The photon's energy was determined via:

The photon's energy $E = h\nu$ (14)

As a result, by putting Eqns. (14) into (15), the mass of the photon may be determined.

Mass of Photon $m = \frac{E}{C^2} = \frac{hv}{c^2}$ (15)

The photon's momentum may be calculated using the following equation. Commencing using relativistic mass:

$$m = \frac{m_0}{\sqrt{1 - \frac{v^2}{c^c}}}$$

We acquire this by squaring both sides of the equation:

$$m^2 \left(1 - \frac{v^2}{c^c}\right) = m_o^2$$

$$m^2 - \frac{m^2 v^2}{c^2} = m_o^2$$

We get this by multiplying both sides of the equation with c^4:

$$m^2 c^4 - m^2 v^2 c^2 = m_o^2 c^4$$

Although,

$m^2 c^4 = E_2 m_o^2 c^4 = E_{02}$, and $m^2 c^2 = p^2$

As a result,

$$E_2 - p^2 c^2 = E_{02}$$

As a result, we can calculate the momentum of every particle using the equation as:

$$p = \frac{\sqrt{E_2 - E_0}}{c}$$

$E_0 = m_0 c^2 = 0$ for a particle with zero rest mass and the momentum of a photon is obtained from the equation.

The photon's momentum $p = E/c$

We may express a photon's momentum in light of its frequency utilizing Eqn. (14):

$p = E/c = h\nu/c$

As $\nu/c = 1/\lambda$, this may be written as:

The Photon's momentum $p = E/c = h\nu/c = h/\lambda$ (16)

Light splits out from an origin in little packets of energy known as photons or quanta, as per this quantum theory of light. Despite being viewed as a particle, the photon's momentum, mass, and energy are defined based on wavelength or frequency, which is purely a wave notion (Clauser, 1974; Derkowska et al., 2000).

As a result, we refer to light as having a double nature. It may be either a wave or a particle; however, never has dual nature at the same moment. To address the question given at the start of this part is light a particle or a wave, light has dual nature. The concept of complementary expresses the dual nature of light: the wave and quantum theories of light complement each other. Light has either a particle or a wave nature depending on the situation, however, never has dual nature at the same moment (Figure 3.8) (Kadrmas et al., 1997; Müller et al., 2017).

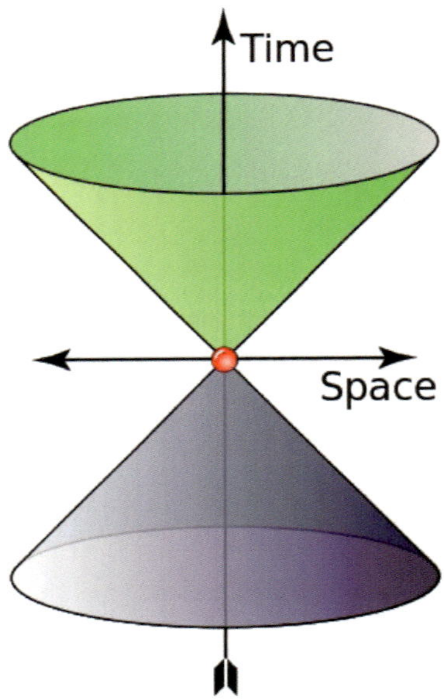

Figure 3.8. The cone depicts the wave 4-vector of a photon's various values. The angular wavenumber (rad·m^{-1}) is represented on the "space" axis, whereas the angular frequency (rad·s^{-1}) is represented on the "time" axis. Green and indigo represent left and right polarization.

Source: https://en.wikipedia.org/wiki/Photon#/media/File:Light_cone_color.svg.

When an electromagnetic wave's wavelength is lengthy, its frequency and, as a result, the energy of the photon (E = $h\nu$) is minimal, therefore we are normally interested in the wave properties of the electromagnetic wave. Tv and radio waves, for instance, have rather longer wavelengths and are typically viewed as waves. The frequency and thus the energy of a photon of an electromagnetic wave is considerable so when the wavelength is short. The electromagnetic wave is thus referred to as a particle. X-rays, for instance, have extremely short wavelengths and are commonly considered particles. This does not rule out the possibility of X-rays acting as waves. Indeed, they do. When X rays are dispersed off a crystal, they act like waves, with the typical diffraction patterns that accompany waves. The crucial point to remember is that light may behave as a particle or a wave, but not both at the same moment (Baek and Kim, 2009; Lee and Kim, 2016).

Let us summarize the photon's characteristics:

Rest Mass $m_0 = 0$ (17)

Energy $E = h\nu$ (18)

Mass $m = E/c^2 = h\nu/c^2$ (19)

The Photon's momentum $p = E/c = h\nu/c = h/\lambda$ (20)

Even though the two cases had been examined for the photons of visible light, it is important to remember that the photon is a particle throughout the electromagnetic spectrum (Kaiser et al., 2009; Greenberg and Sultanov, 2017).

3.5. THE COMPTON EFFECT

If light may act like a particle, the photon, why do not think of a photon colliding with a free electron like a collision between two billiard balls? In honor to Arthur Holly Compton, quite a collision between a free electron and a photon is known as Compton scattering or the *Compton effect* (1892–1962). X rays have been employed to generate a large photon for the collision. (Remember that X rays contain a high frequency, which means that their energy, E = $h\nu$, is considerable, and their mass, m = E/ c^2, is likewise large.) A target composed of carbon is utilized to obtain a free electron (Tanaka et al., 2015; Shakhmuratov et al., 2020). The valence electron of the carbon atom is quite weakly bonded; therefore, the electron seems to be a free electron when contrasted to the photon's starting energy. As seen in Figure 3.9, the collision between the electron and photon may be visualized. The electron is first at rest, and the incoming photon possesses energy (E = $h\nu$) and momentum (p = E/c) (Figure 3.9) (Legero et al., 2004; Hubbard et al., 2007).

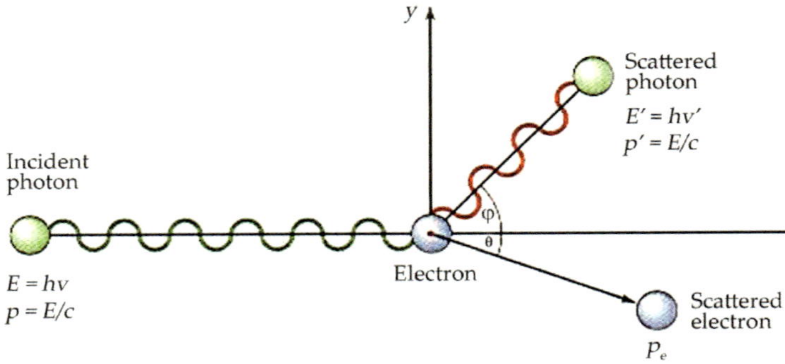

Figure 3.9. Compton scattering.

Source: https://www.mdpi.com/2073-4352/11/5/525/htm.

The electron is discovered to be dispersed at an angle from the photon's initial path after the impact. Since the electron shifted after the collision, it should have received considerable energy (Sultanov and Greenberg, 2017). But wherein did this energy originate?

It was most likely caused by the incident photon. If this is correct, the dispersed photon should have less energy than the incoming photon, and hence its wavelength should be shorter and have been altered as well (Baek and Kim, 2007; Cheng, 2019). Let us name the dispersed photon's energy E,' wherein;

E' = $h\nu$'

As a result, its final momentum is:

p' = E'/c = $h\nu$'/c

Conservation of momentum applies to the collision in Figure 3.9 as momentum is conserved throughout all collisions. Furthermore, keep in mind that the collision is 2-dimensional. The x-element of momentum and the y-element of momentum should both be sustained since the vector momentum is retained (Bonnor, 1973; Altewischer et al., 2002). For the x-element of momentum, the law of conservation may be expressed as:

$p\mathbf{p} + 0 = p\mathrm{p}' \cos \phi + p\mathbf{e} \cos \theta$

and for the *y*-element,

$0 + 0 = pp' \sin \phi - p\mathbf{e} \sin \theta$

wherein; $p\mathbf{e}$ is the scattered electron's momentum, pp' is the scattered photon's momentum, and $p\mathbf{p}$ is the originating photon's momentum. When the photon's energy and momentum are substituted, the equations become:

$h\nu/c = h\nu'/c \cos \phi + p\mathbf{e} \cos \theta$

$0 = h\nu'/c \sin \phi - p\mathbf{e} \sin \theta$

We cannot manage all of the unknowns (ν,' θ, ϕ, $p\mathbf{e}$) right now, so let us remove θ from such two equations via squaring, rearranging, and adding them.

i.e.,

$$p\mathbf{e} \cos \theta = \frac{h\nu}{c} - \frac{h\nu'}{c} \cos \phi$$

$$p\mathbf{e} \sin \theta = \frac{h\nu'}{c} \sin \phi$$

$$pe2 \cos 2\theta = \frac{(h\nu)2}{c2} - \frac{2(h\nu)(h\nu')\cos}{c2} + \frac{(h\nu')2}{c2} \cos 2\phi$$

$$pe2 \sin 2\theta = \frac{(h\nu')2}{c2} \sin 2\phi$$

$$pe2 (\sin 2\theta + \cos 2\theta) = \frac{(h\nu)2}{c2} + \frac{(h\nu')2}{c2} (\sin 2\phi + \cos 2\phi) - \frac{2(h\nu)(h\nu')}{c2} \cos \phi$$

Although sin2θ + cos2θ = 1, we obtain:

$$pe2 = \frac{(h\nu)2}{c2} + \frac{(h\nu')2}{c2} - \frac{2(h\nu)(h\nu')}{c2} \cos \phi \qquad (21)$$

As a result, the angle θ is removed from the equation. Let us seek a solution to get rid of the pe, or electron momentum (Evans, 1958; Regelman et al., 2001). We may resolve for $pe2$ by squaring Eqn. (21) and we get:

$$pe2 = \frac{Ee2 - E0e2}{c2} \qquad (22)$$

However, according to Eqn. (22), the entire energy of the electron Ee is:

$E\mathbf{e} = \mathrm{KE}\mathbf{e} + E\mathbf{0e}$

wherein, KEe is the electron's kinetic energy; and E0e has its rest mass. When Eqn. (21) is substituted back into Eqn. (22), we get:

$$pe2 = \frac{(\text{KE}\mathbf{e} + E0\mathbf{e})\,2 - E0\text{e}2}{c2}$$

$$= \frac{\text{KEe}2 + 2E0\mathbf{e}\,\text{KE}\mathbf{e} + E02 - E02}{c2}$$

$$pe2 = \frac{\text{KEe}2 + 2E0\mathbf{e}\,\text{Ke}\mathbf{e}2}{c2} \tag{23}$$

However, when we apply the rule of energy conservation to the collision in Figure 3.9, we obtain:

$$E = E' + \text{KE}\mathbf{e}$$

$$h\nu = h\nu' + \text{KEe} \tag{24}$$

wherein; E' is the energy of the dispersed photon; E is the entire energy of the system, and KEe is the kinetic energy given to the electron throughout the collision (Raman and Krishnan, 1928; Everett et al., 1977). As a result of Eqn. (24), the electron's KE is:

$$\text{KE}\mathbf{e} = h\nu - h\nu' \qquad (25)$$

By putting the KE value from Eqn. (25) and $E\mathbf{0e} = m_0c^2$, the electron's rest energy, back into Eqn. (23), we obtain, for the electron's momentum:

$$pe2 = \frac{(h\nu - h\nu')2 + 2m0c^2(h\nu - h\nu')}{c2}$$

$$pe2 = \frac{(h\nu)2}{c2} + \frac{(h\nu')2}{c2} - \frac{2h\nu h\nu'}{c2} + 2m_0(h\nu - h\nu')$$

We may equal Eqns. (21) and (23) to remove pe since we now have two distinct equations for the electron's momentum. Thus,

$$\frac{(h\nu)2}{c^2} + \frac{(h\nu')2}{c^2} - \frac{2h\nu h\nu'}{c^2} + 2m_0(h\nu - h\nu') = \frac{(h\nu)2}{c^2} + \frac{(h\nu')2}{c^2} - \frac{2h\nu h\nu'}{c^2}\cos\phi$$

By making it simple we obtain:

$$2m_0(h\nu - h\nu') = \frac{2h\nu h\nu'}{c^2} - \frac{2(h\nu)(h\nu')}{c^2}\cos\phi$$

$$h\nu - h\nu' = \frac{h\nu h\nu'}{m0c^2}(1 - \cos\phi)$$

$$\frac{\nu - \nu'}{\nu\nu'} = \frac{h}{m0c^2}(1 - \cos\phi)$$

Though, since $\nu = c/\lambda$ we get:

$$\frac{c/-c/'}{(c/)-(c/')} = \frac{h}{m0c^2}(1-\cos\phi) \tag{26}$$

Eqn. (26) shows how the wavelength of the dispersed photon changes as a function of the scattering angle ϕ. The Compton wavelength is defined as $h = 2.426 \times 10^{-12}$ m = 0.002426 nm which has the dimension of length. When an energetic photon collides with an electron, the dispersed light has a wavelength that differs from the input light. In 1923, A. H. Compton, who got the Nobel Prize in 1927 for his work, verified the changed wavelength of the scattered photon (Mandl and Skyrme, 1952; Antipov et al., 1983). In a real experiment, the dispersed photons contain both the incoming and modified wavelengths. Since certain incident photons have been dispersed by the atom, the incident wavelength may be detected in the dispersed photons. The rest mass of the electron m_0 should be substituted by the mass M of the complete atom in Eqn. (26) in this situation. The Compton wavelength h/MC is so tiny that the shift in ratio for such photons is very modest to notice since M is very bigger than m_0. As a result, the dispersed photons have a similar wavelength (Figure 3.10) (Dirac, 1927; Weymann, 1965).

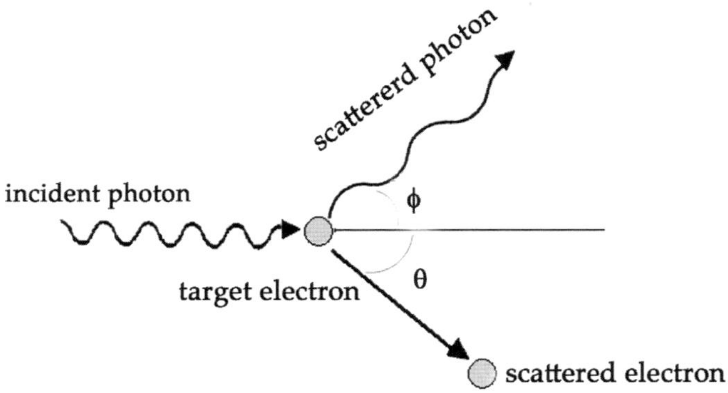

Figure 3.10. The directions of the dispersed photon and electron for the direction of the incoming photon are shown in the geometry of Compton scattering.

Source: https://www.researchgate.net/figure/The-geometry-of-Compton-scattering-showing-the-directions-of-the-scattered-photon-and_fig1_236737231.

3.6. THE WAVE NATURE OF PARTICLES

We have seen that light has two natures: it operates as a particle and as a wave. In his PhD dissertation in 1924, the French physicist Louis de Broglie (1892–1987) postulated that particles must have a wave feature, supposing symmetry in nature. Since photon momentum had been demonstrated to be:

$$p = \frac{h}{\lambda} \qquad (27)$$

The wavelength of the wave linked with a particle having momentum p, according to Louis de Broglie, must be provided as:

$$\lambda = \frac{h}{p} \qquad (28)$$

The de Broglie connection is the name given to Eqn. (26). As a result, de Broglie hypothesized that the wave-particle dual nature that governs electromagnetic waves must also govern particles. As a result, an electron may be thought of as both a wave and a particle. Rather than resolving the issue of electromagnetic waves' wave-particle dual nature, de Broglie expanded it to encompass matter (Breit, 1926; Vokos et al., 2000). Since Planck's invariant "*h*" is so little, the wave nature of a particle does not reveal itself until the particle's mass m is likewise very limited, as seen in instance 3.8. That is why the wave character of particles is not something that we are familiar with (Strikman and Frankfurt, 1980; Gareev et al., 1997).

When L. H. Germer and C. J. Davisson demonstrated that electrons might be diffracted via crystal in 1927, de Broglie's theory had been instantly validated. Simultaneously, G. P. Thomson conducted an independent experiment, dispersion of electrons from extremely thin metal foils and obtaining the conventional diffraction patterns linked with waves. Since then, diffraction patterns are recorded with neutrons, protons, helium atoms, and hydrogen atoms, providing conclusive proof for particle-wave nature. De Broglie was awarded the Nobel Prize in Physics in 1929 for his research on the duality of particles. In 1937, Thomson, and Davisson won a Nobel Prize for their experimental proof of particle-wave nature (Figure 3.11) (Ravichandran et al., 2014).

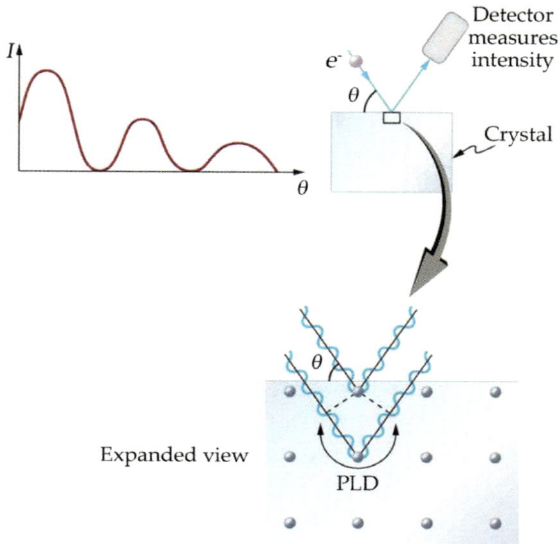

Figure 3.11. The upper left diffraction pattern is created by dispersing electrons from a crystal and is graphed as a function of incidence angle compared to the regular array of atoms in a crystal, as seen at the bottom. Electrons dispersed from atoms in the 2nd layer go further than those dispersed from atoms in the upper layer. There is constructive interference when the path length difference (PLD) is an integral wavelength.

Source: https://courses.lumenlearning.com/physics/chapter/29-6-the-wave-nature-of-matter/.

3.7. THE WAVE REPRESENTATION OF A PARTICLE

A particle may be expressed as a wave, as we have just seen. An electromagnetic wave was coupled with a photon. What type of wave, however, is linked with a particle? It is not an electromagnetic wave, for sure. The wave was dubbed a pilot wave by De Broglie since he felt it guided the particle throughout its journey. To emphasize their connection to matter, the waves are known as matter waves (Hu et al., 2000; Anglin, 2010). The wave is now simply known as the wave function and is depicted by Ψ .

We argue that the wave function's value is connected to the probability of detecting the particle at a certain location and time since it refers to the velocity of a particle. The probability P something might be anywhere at a specific moment might range from zero to one. If the probability P is equal to zero, then the particle is guaranteed to be missing. If the probability P is equal to one, then the particle is guaranteed to be present. If the probability P is between zero and one, then the likelihood of detecting the particle there is that value (Dodonov and Andreata, 2000; Tu, 2019). If the probability P is equal to 0.20, then there is a 20% chance of detecting the particle at the provided location and time. The wave function may not describe the probability of detecting the particle at a particular time and location since the amplitude of each wave swings between +ve and –ve values. Therefore, the probability density is the amount Ψ2 that is always +ve. The probability density Ψ2 represents the likelihood of detecting the particle in the plane (x, y, z) at time t.

The modern science of wave mechanics, or quantum mechanics as it was later termed, is concerned with finding the wave function of every particle or set of particles (Figure 3.12) (Dolce and Perali, 2014; Mahajan, 2019).

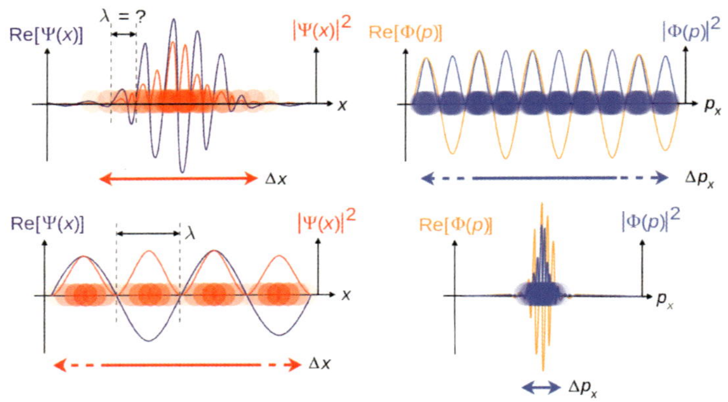

Figure 3.12. Wavefunctions representing quantum particles' position 'x' and momentum 'p.' The probability density of locating a particle having position 'x' or velocity component 'p' correlates to the color opacity of the particles.

Source: https://en.wikipedia.org/wiki/Wave%E2%80%93particle_duality.

What is the best way to represent a particle with a wave? Remember from basic physics that the function defines a wave traveling to the right.

$y = A \sin(kx - \omega t)$

Whereas the wave number k is:

$$k = \frac{2\pi}{\lambda}$$

and the angular frequency ω is expressed as:

$\omega = 2\pi f$

or as $f = \nu$

so,

$\omega = 2\pi\nu$

Also, remember that the wave's velocity is expressed as:

$$v = \frac{\omega}{k}$$

As a result, we would start our examination of matter waves by attempting to explain the wave function:

$\Psi = A \sin(kx - \omega t)$ (29)

Figure 3.13 depicts a plot of the wave function at t = 0 for this function (a). The 1st thing to notice about this image is that the wave is too diffuse to depict a particle. Recall that the particle should be located inside the wave. Since the wave expanded to infinity, the location of the particle is arbitrary (Figure 3.13) (Miller et al., 2009; Eibenberger et al., 2013).

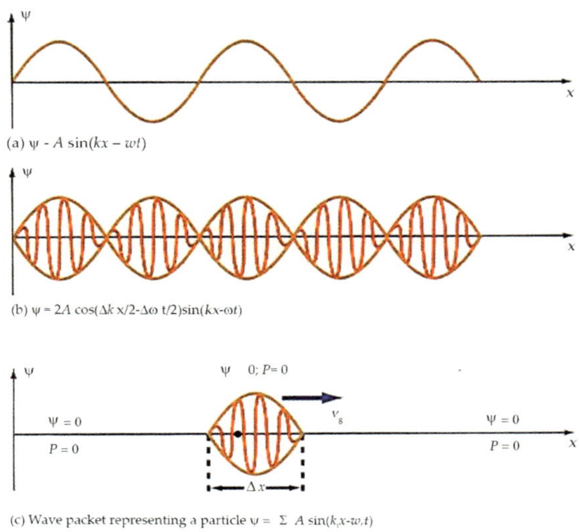

Figure 3.13. Particle depiction as a wave.

Source: https://www.researchgate.net/figure/A-wave-packet-representation-of-a-material-particle_fig8_233327991.

Since one of the properties of waves would be that they follow the principle of superposition, a wave depiction may be discovered by combining various waves. Let us take two waves with slightly distinct wave numbers and angular frequencies as an instance (Lax, 1951; Shpenkov and Kreidik, 2004). Take, for example, the two waves:

$\Psi_1 = A \sin(k_1 x - \omega_1 t)$

$\Psi_2 = A \sin(k_2 x - \omega_2 t)$

Whereas,

$k_2 = k_1 + \Delta k$

and,

$\omega_2 = \omega_1 + \Delta\omega$

The result of adding these two waves is:

$$\Psi = \Psi_1 + \Psi_2$$
$$= A \sin(k_1 x - \omega_1 t) + A \sin(k_2 x - \omega_2 t)$$

Appendix B illustrates the addition of two sine waves, which are intended to represent:

$$\text{sinBsinC} = 2\sin(\frac{B+C}{2})\cos(\frac{B-C}{2})$$

let,

$B = k_1 x - \omega_1 t$

and,

$C = k_2 x - \omega_2 t$

we obtain:

$$\Psi = 2A\sin\frac{k1x - 1t + k2x - 2t}{2}\cos\frac{k1x - 1t - k2x + 2t}{2}$$
$$= 2A\sin(\frac{k1x - 1t + (k1 + k)x - (1+)t}{2})\cos(\frac{k1x - 1t - (k1 + k)x - (1+)t}{2})$$
$$= 2A\sin(\frac{2kx + (k)x - 2t - ()t}{2})\cos(\frac{-k}{2}x + \frac{}{2}t)$$

To illustrate the general case, we removed subscript 1 from k and ω. So, as a rough estimate:

$2kx + (\Delta k)x \approx 2kx$

and,

$-2\omega t - (\Delta\omega)t \approx -2\omega t$

Thus,

$$\Psi = 2A\sin(kx - t)\cos\left(\frac{-k}{2}x + \frac{}{2}t\right) \quad (30)$$

One of the cosine function's features is that cos(−θ) = cos θ. The wave function is transformed with this relationship.

$$\Psi = 2A\cos\left(\frac{k}{2}x - \frac{}{2}t\right)\sin(kx - t) \quad (31)$$

Figure 3.13(b) depicts a visualization of Eqn. (31). The modulated amplitude of this wave is provided by the 1st section of Eqn. (31) as:

$$A_m = 2A\cos\left(\frac{k}{2}x - \frac{}{2}t\right)$$

Such superposition of waves provides a more accurate picture of a particle. Every segment of the wave that is modified depicts a group of waves, and every group might depict a particle. Particle velocity is represented by the velocity of the group of waves (Salehi and Bisabr, 2003; Jain, 2004).

Figure 3.13(b) and Eqn. (31) approximate a wave depiction of the particle. If an unlimited quantity of waves that vary somewhat in wavenumber and angular frequency had been combined, the wave function will result.

$$\Psi = \sum_{i=1}^{\infty} 2A\sin(k1x - 1t)$$

This is referred to as a wave packet, as seen in Figure 3.13(c). This wave packet can represent the motion of a particle. Because the wave function Ψ is zero outside of the packet, the probability of finding the particle is likewise zero outside of the packet. The wave packet bonds exactly to the particle inside the region Δx illustrated in Figure 3.13(c), and the wave packet moves at the same speed as the waves' group velocity. Quantum mechanics, often known as wave mechanics, is concerned with determining the wave function associated with a particle or particle system.

3.8 THE HEISENBERG UNCERTAINTY PRINCIPLE

A basic constraint in the precision of measuring a particle's location and momentum is among the features of the dual nature of matter. The modulated wave is shown in Figure 3.13(b), which is duplicated in Figure 3.13, shows it in a very simple form. In the 1st group of the modulated wave, a particle is visible. The particle's exact location is unknown as it is included within the packet of the wave. Its positional uncertainty is limited to Δx, the breadth of the full-wave packet or wave group. Figure 3.13 shows the wavelength of the modulated amplitude Δ**m**, and we may see that a wave group is just half that distance. As a result, the uncertainty in the particle's position is provided by:

$$\Delta x = \frac{\mathbf{m}}{2} \lambda \mathbf{m} \tag{30}$$

The de Broglie relation, Eqn. (28), for momentum may be used to determine the uncertainty in the momentum (Figure 3.14).

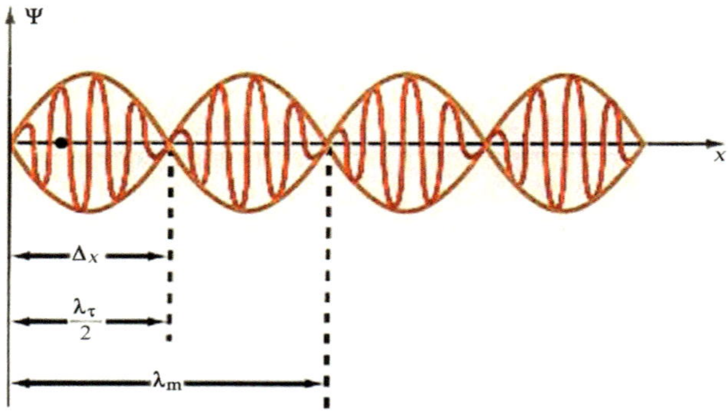

Figure 3.14. A particle's location and momentum are limited.

Source: http://abyss.uoregon.edu/~js/21st_century_science/lectures/lec14.html.

$$p = \frac{h}{\lambda} \tag{31}$$

and since the wavelength is expressed as the wavenumber,

$$\lambda = \frac{2}{k} \tag{32}$$

By putting the Eqns. (32) into (31) then we obtain:

$$p = \frac{h}{\lambda} = \frac{h}{2\pi/k} = \frac{h\,k}{2\pi} \tag{33}$$

The uncertainty in momentum, as determined by Eqn. (33), is:

$$\Delta p = \frac{h}{2}\Delta k \qquad (34)$$

There may be a Δk associated with the wave packet since it is composed of several waves. This indicates that when a particle is represented as a wave, the wavenumber k is inherently unpredictable, which indicates that the particle's momentum is also unpredictable. The wavenumber of the modulated wave $\Delta k\mathbf{m}$ is calculated for the particular example shown in Figure 3.14.

$A\mathbf{m} = 2A\cos(k\mathbf{m}x - \omega\mathbf{m}t)$

and from Eqn. (32) we get:

$$k\mathbf{m} = \frac{k}{2} \qquad (35)$$

However, according to the definition of a wave number,

$$k\mathbf{m} = \frac{k}{2}$$

$$k\mathbf{m} = \frac{2}{\mathbf{m}} \qquad (36)$$

By putting Eqns. (36) into (35) then we get, for Δk,

$$k = 2k\mathbf{m} = \mathbf{2}\,(\frac{2}{\mathbf{m}})$$

By putting the uncertainty for Δk, Eqn. (34), into the uncertainty for Δp, Eqn. (34), we obtain:

$$\Delta p = \frac{h}{2}k = \frac{h}{}\,2(\frac{2}{\mathbf{m}}) = \frac{h}{\boldsymbol{hm}/2}$$

The uncertainty between a particle's location and momentum is found by putting Eqn. (36) for $\lambda\mathbf{m}/2$ into Eqn. (37).

$$\Delta p = \frac{\mathrm{h}}{x}$$

or,

$$\Delta p\Delta x = h \qquad (37)$$

Since the lowest uncertainties that a particle may have are Δx and Δp, its values are generally more than this, and its product has been typically higher than the value of h. To demonstrate this, Eqn. (37) is frequently written with an inequality sign, as in:

$\Delta p \Delta x \geq h$

Figure 3.13(b) shows a modulated wave that substantially improved the analysis of the wave packet. Figure 3.13(c) depicts the result of a more advanced analysis performed on the more acceptable wave packet.

$\Delta p \Delta x \geq$ (38)

Whereas the symbol h, known as h bar, is:

$$\mathrm{h} = \frac{h}{2} = 1.05 \times 10^{-54}$$

The Heisenberg uncertainty principle is the equation. It states that a particle's location and momentum may not be accurately determined at the same time. Every measurement carries a specific amount of inherent uncertainty. The measurement device has no impact on this uncertainty. It is the result of matter's dual nature of wave-particle (Friese et al., 1998).

3.9. DIFFERENT FORMS OF THE UNCERTAINTY PRINCIPLE

There is a limit to the number of simultaneous measurements that can be done for a particle's location and momentum, as well as for its angular momentum and angular position, as well as for the particle's energy, and for the time during which the energy measurement is performed.

One of the definitions of the angular momentum of a particle is:

$L = rp\sin\theta$ (39)

The velocity, and thus the momentum, of a particle traveling in a circle of radius r, is perpendicular to the radius. As a result, $\theta = 90°$ and $\sin 90° = 1$. As a result, the angular momentum of a particle may be written as the multiplication of the particle's linear momentum and the circle's radius. In other words,

$L = rp$ (40)

We may examine the influence of the uncertainty principle on a particle in rotary movement using this notion of angular momentum (Yukawa,

1935). Whenever a particle goes across the angle θ, we obtain x, which is the displacement of the particle alongside the arc of the circle.

$$x = r\theta$$

When the uncertainty Δx is expressed in terms of the uncertainty Δθ in angle, the result is:

$$\Delta x = r\Delta\theta$$

We obtain the following result when we substitute this uncertainty in the relation to Heisenberg's uncertainty:

$$\Delta x \Delta p \geq h$$

$$r\Delta\theta\Delta p \geq h$$

$$(\Delta\theta)(r\Delta p) \geq h \qquad (41)$$

However, Eqn. (41), which provided the angular momentum of the particle, also provides the uncertainty associated with this angular momentum.

$$\Delta L = r\Delta p \qquad (42)$$

However, this is precisely one of the components in Eqn. (41). Putting Eqns. (42) into (41) obtains the Heisenberg uncertainty principle for rotary motion as:

$$\Delta\theta\ \Delta L \geq h \qquad (43)$$

In this formulation of Heisenberg's uncertainty principle, the multiplication of the particle's uncertainty in angular momentum and angular position has always been equal to or higher than the value *h*. Therefore, if the angular location of a particle is precisely known, Δθ = 0, the uncertainty in its angular momentum is unlimited. Alternatively, if the angular momentum is precisely known, ΔL = 0, we have no notion of which particle is in the circle. The link between the uncertainty of a particle's energy and the uncertainty of its measurement time is as below. Since the velocity of a particle is provided by v = Δx/Δt, the distance traveled by the particle while the measurement is:

$$\Delta x = v\Delta t \qquad (44)$$

The de Broglie relation calculates the particle's momentum as follows:

$$p = \frac{h}{\lambda} = \frac{h\nu}{v} = \frac{E}{v} \qquad (45)$$

Since hν = E and 1/λ = ν/v. Derived from Eqn. (45), the uncertainty of momentum in terms of the uncertainty of its energy is:

$$\Delta p = \frac{E}{v} \tag{46}$$

Eqns. (44) and (46) are substituted into the relation of Heisenberg uncertainty.

We obtain:

$$\Delta x \Delta p \geq h$$

$$\Delta E \Delta t \geq h \tag{47}$$

According to Eqn. (47), the product of the uncertainty in measuring a particle's energy and the uncertainty in measuring the particle's time has always been equal to or higher than h. As a result, measuring the energy of a particle precisely, $\Delta E = 0$, will require an indefinite amount of time. If we evaluate the particle at $\Delta t = 0$, we would have no notion of how much energy the particle contains (ΔE will be infinite).

REFERENCES

1. Acín, A., & Masanes, L., (2016). Certified randomness in quantum physics. *Nature, 540*(7632), 213–219.
2. Adawi, I., (1964). Theory of the surface photoelectric effect for one and two photons. *Physical Review, 134*(3A), A788.
3. Agostini, P., & Petite, G., (1988). Photoelectric effect under strong irradiation. *Contemporary Physics, 29*(1), 57–77.
4. Altewischer, E., Van, E. M. P., & Woerdman, J. P., (2002). Plasmon-assisted transmission of entangled photons. *Nature, 418*(6895), 304–306.
5. Anglin, J., (2010). Particles of light. *Nature, 468*(7323), 517, 518.
6. Antipov, Y. M., Batarin, V. A., Bessubov, V. A., Budanov, N. P., Gorin, Y. P., Denisov, S. P., & Travkin, V. I., (1983). Measurement of π^--meson polarizability in pion Compton effect. *Physics Letters B, 121*(6), 445–448.
7. Arndt, M., Juffmann, T., & Vedral, V., (2009). Quantum physics meets biology. *HFSP Journal, 3*(6), 386–400.
8. Arutyunian, F. R., & Tumanian, V. A., (1963). The Compton effect on relativistic electrons and the possibility of obtaining high energy beams. *Physics Letters, 4*(3), 176–178.
9. Baek, S. Y., & Kim, Y. H., (2007). Spectral properties of entangled photons generated via type-I spontaneous parametric downconversion. In: *Conference on Lasers and Electro-Optics* (p. JTuA2). Optical Society of America.
10. Baek, S. Y., & Kim, Y. H., (2009). Spectral properties of entangled photons generated via type-I frequency-nondegenerate spontaneous parametric down-conversion. *Physical Review A, 80*(3), 033814.
11. Barenco, A., (1996). Quantum physics and computers. *Contemporary Physics, 37*(5), 375–389.
12. Bonnor, W. B., (1973). Properties of charged photons. *Il Nuovo Cimento A (1965–1970), 15*(4), 767–784.
13. Boyer, T. H., (1969). Derivation of the blackbody radiation spectrum without quantum assumptions. *Physical Review, 182*(5), 1374.
14. Boyer, T. H., (2018). Blackbody radiation in classical physics: A historical perspective. *American Journal of Physics, 86*(7), 495–509.

15. Breit, G., (1926). A correspondence principle in the Compton effect. *Physical Review, 27*(4), 362.
16. Brout, R., Massar, S., Parentani, R., & Spindel, P., (1995). A primer for black hole quantum physics. *Physics Reports, 260*(6), 329–446.
17. Busch, P., Grabowski, M., & Lahti, P. J., (1997). *Operational Quantum Physics* (Vol. 31). Springer Science & Business Media.
18. Chang, S. L., & Rhee, K. T., (1984). Blackbody radiation functions. *International Communications in Heat and Mass Transfer, 11*(5), 451–455.
19. Cheng, Z., (2019). Thermodynamic properties of coherent photons in self-focusing nonlinear waveguides. *Journal of Nonlinear Optical Physics & Materials, 28*(01), 1950006.
20. Clauser, J. F., (1974). Experimental distinction between the quantum and classical field-theoretic predictions for the photoelectric effect. *Physical Review D, 9*(4), 853.
21. Cooke, W. E., & Gallagher, T. F., (1980). Effects of blackbody radiation on highly excited atoms. *Physical Review A, 21*(2), 588.
22. Cornelius, C. M., & Dowling, J. P., (1999). Modification of Planck blackbody radiation by photonic band-gap structures. *Physical Review A, 59*(6), 4736.
23. Derkowska, B., Sahraoui, B., Phu, X. N., Glowacki, G., & Bala, W., (2000). Study of linear optical properties and two-photons absorption in Zn1− xMgxSe thin layers. *Optical Materials, 15*(3), 199–203.
24. Deutsch, D., & Lockwood, M., (1994). The quantum physics of time travel. *Scientific American, 270*(3), 68–74.
25. Deutsch, D., Ekert, A., & Lupacchini, R., (2000). Machines, logic and quantum physics. *Bulletin of Symbolic Logic, 6*(3), 265–283.
26. Dicke, R. H., Peebles, P. J. E., Roll, P. G., & Wilkinson, D. T., (1965). Cosmic black-body radiation. *The Astrophysical Journal, 142*, 414–419.
27. Dirac, P. A., (1927). The Compton effect in wave mechanics. In: *Mathematical Proceedings of the Cambridge Philosophical Society* (Vol. 23, No. 5, pp. 500–507). Cambridge University Press.
28. Dodonov, V. V., & Andreata, M. A., (2000). Deflection of quantum particles by impenetrable boundary. *Physics Letters A, 275*(3), 173–181.

29. Dolce, D., & Perali, A., (2014). *Probing the Geometric Nature of Particles Mass in Graphene Systems.* arXiv preprint arXiv:1403.7037.
30. Dunbar, R. C., & McMahon, T. B., (1998). Activation of unimolecular reactions by ambient blackbody radiation. *Science, 279*(5348), 194–197.
31. Economou, E. N., (2006). *Green's Functions in Quantum Physics* (Vol. 7). Springer Science & Business Media.
32. Eibenberger, S., Gerlich, S., Arndt, M., Mayor, M., & Tüxen, J., (2013). Matter-wave interference of particles selected from a molecular library with masses exceeding 10000 amu. *Physical Chemistry Chemical Physics, 15*(35), 14696–14700.
33. Eisberg, R., & Resnick, R., (1985). *Quantum Physics of Atoms, Molecules, Solids, Nuclei, and Particles,* 864.
34. Evans, R. D., (1958). Compton effect. In: *Corpuscles and Radiation in Matter II/Korpuskeln und Strahlung in Materie II* (pp. 218–298). Springer, Berlin, Heidelberg.
35. Everett, D. B., Fleming, J. S., Todd, R. W., & Nightingale, J. M., (1977). Gamma-radiation imaging system based on the Compton effect. In: *Proceedings of the Institution of Electrical Engineers* (Vol. 124, No. 11, pp. 995–1000). IET Digital Library.
36. Ford, G. W., Lewis, J. T., & O'connell, R. F., (1985). Quantum oscillator in a blackbody radiation field. *Physical Review Letters, 55*(21), 2273.
37. French, S., & Redhead, M., (1988). Quantum physics and the identity of indiscernibles. *The British Journal for the Philosophy of Science, 39*(2), 233–246.
38. French, S., (1989). Identity and individuality in classical and quantum physics. *Australasian Journal of Philosophy, 67*(4), 432–446.
39. Friese, M. E., Nieminen, T. A., Heckenberg, N. R., & Rubinsztein-Dunlop, H., (1998). Optical alignment and spinning of laser-trapped microscopic particles. *Nature, 394*(6691), 348–350.
40. Gallagher, T. F., & Cooke, W. E., (1979). Interactions of blackbody radiation with atoms. *Physical Review Letters, 42*(13), 835.
41. Gareev, F. A., Barabanov, M. Y., & Kazacha, G. S., (1997). Wave nature of particles. *Dihadronic and Dileptonic Resonances* (No. JINR-E—3-97-213).
42. Gazeau, J. P., (2009). *Coherent states in Quantum Physics* (p. 344). Berlin: Wiley-VCH.

43. Geiger, J. A., (2018). Measurement quantization unites classical and quantum physics. *Journal of High Energy Physics, Gravitation and Cosmology, 4*(2), 262.
44. Gel'fand, I. M., & Yaglom, A. M., (1960). Integration in functional spaces and its applications in quantum physics. *Journal of Mathematical Physics, 1*(1), 48–69.
45. Gerlach, U. H., (1976). The mechanism of blackbody radiation from an incipient black hole. *Physical Review D, 14*(6), 1479.
46. Giamarchi, T., (2003). *Quantum Physics in One Dimension* (Vol. 121). Clarendon Press.
47. Greenberg, Y. S., & Sultanov, A. N. E., (2017). Influence of the nonradiative decay of qubits into a common channel on the transport properties of microwave photons. *JETP Letters, 106*(6), 406–410.
48. Greenberger, D., Hentschel, K., & Weinert, F., (2009). *Compendium of Quantum Physics* (Vol. 3, pp. 29–42). Berlin, Heidelberg: Springer Berlin Heidelberg.
49. Greenstein, G., & Hartke, G. J., (1983). Pulselike character of blackbody radiation from neutron stars. *The Astrophysical Journal, 271*, 283–293.
50. Guhr, T., Müller–Groeling, A., & Weidenmüller, H. A., (1998). Random-matrix theories in quantum physics: Common concepts. *Physics Reports, 299*(4–6), 189–425.
51. Haslinger, P., Jaffe, M., Xu, V., Schwartz, O., Sonnleitner, M., Ritsch-Marte, M., & Müller, H., (2018). Attractive force on atoms due to blackbody radiation. *Nature Physics, 14*(3), 257–260.
52. Ho, I., Pucci, G., Oza, A. U., & Harris, D. M., (2021). *Capillary Surfers: Wave-Driven Particles at a Fluid Interface.* arXiv preprint arXiv:2102.11694.
53. Holevo, A. S., (1973). Statistical problems in quantum physics. In: *Proceedings of the Second Japan-USSR Symposium on Probability Theory* (pp. 104–119). Springer, Berlin, Heidelberg.
54. Hu, W., Barkana, R., & Gruzinov, A., (2000). Fuzzy cold dark matter: The wave properties of ultralight particles. *Physical Review Letters, 85*(6), 1158.
55. Hubbard, R., Ovchinnikov, Y. B., Cheung, J., Fletcher, N., Murray, R., & Sinclair, A. G., (2007). Measurements of statistical properties of single photons emitted by a solitary NV center in synthetic diamond. *Journal of Modern Optics, 54*(2, 3), 441–451.

56. Jain, Y. S., (2004). Wave mechanics of two hardcore quantum particles in A 1-D box. *Central European Journal of Physics, 2*(4), 709–719.
57. Kadrmas, D. J., Frey, E. C., & Tsui, B. M., (1997). Analysis of the reconstructibility and noise properties of scattered photons in Tc SPECT. *Physics in Medicine & Biology, 42*(12), 2493.
58. Kaiser, F., Jacques, V., Batalov, A., Siyushev, P., Jelezko, F., & Wrachtrup, J., (2009). *Polarization Properties of Single Photons Emitted by Nitrogen-Vacancy Defect in Diamond at Low Temperature*. arXiv preprint arXiv:0906.3426.
59. Kelly, R. E., (1981). Thermodynamics of blackbody radiation. *American Journal of Physics, 49*(8), 714–719.
60. Klassen, S., (2011). The photoelectric effect: Reconstructing the story for the physics classroom. *Science & Education, 20*(7), 719–731.
61. Lamb, Jr. W. E., & Scully, M. O., (1968). *The Photoelectric Effect Without Photons* (No. TS-QED-68-1).
62. Lax, M., (1951). Multiple scattering of waves. *Reviews of Modern Physics, 23*(4), 287.
63. Lee, J. C., & Kim, Y. H., (2016). Spatial and spectral properties of entangled photons from spontaneous parametric down-conversion with a focused pump. *Optics Communications, 366*, 442–450.
64. Legero, T., Wilk, T., Hennrich, M., Rempe, G., & Kuhn, A., (2004). Quantum beat of two single photons. *Physical Review Letters, 93*(7), 070503.
65. Leonhardt, U., (2003). Quantum physics of simple optical instruments. *Reports on Progress in Physics, 66*(7), 1207.
66. Mahajan, S. M., (2019). *Resonant Energization of Relativistic Particles by an Intense Electromagnetic Wave*. arXiv preprint arXiv:1910.08147.
67. Mandl, F., & Skyrme, T. H. R., (1952). The theory of the double Compton effect. *Proceedings of the Royal Society of London. Series A. Mathematical and Physical Sciences, 215*(1123), 497–507.
68. McKagan, S. B., Handley, W., Perkins, K. K., & Wieman, C. E., (2009). A research-based curriculum for teaching the photoelectric effect. *American Journal of Physics, 77*(1), 87–94.
69. Michelini, M., Ragazzon, R., Santi, L., & Stefanel, A., (2000). Proposal for quantum physics in secondary school. *Physics Education, 35*(6), 406.

70. Middelmann, T., Falke, S., Lisdat, C., & Sterr, U., (2012). High accuracy correction of blackbody radiation shift in an optical lattice clock. *Physical Review Letters, 109*(26), 263004.

71. Miller, D. L., Kubista, K. D., Ruan, M., De Heer, W. A., First, P. N., Rutter, G. M., & Stroscio, J. A., (2009). Quantization of zero-mass particles in graphene. *Science, 324*, 927–929.

72. Morandi, G., (1992). *The Role of Topology in Classical and Quantum Physics* (Vol. 7). Springer Science & Business Media.

73. Müller, P., Tentrup, T., Bienert, M., Morigi, G., & Eschner, J., (2017). Spectral properties of single photons from quantum emitters. *Physical Review A, 96*(2), 023861.

74. Ossiander, M., Riemensberger, J., Neppl, S., Mittermair, M., Schäffer, M., Duensing, A., & Kienberger, R., (2018). Absolute timing of the photoelectric effect. *Nature, 561*(7723), 374–377.

75. Piron, C., (1976). On the foundations of quantum physics. In: *Quantum Mechanics, Determinism, Causality, and Particles* (pp. 105–116). Springer, Dordrecht.

76. Popper, K. R., (1950). Indeterminism in quantum physics and in classical physics: Part II. *The British Journal for the Philosophy of Science, 1*(3), 173–195.

77. Pratt, R. H., Ron, A., & Tseng, H. K., (1973). Atomic photoelectric effect above 10 keV. *Reviews of Modern Physics, 45*(2), 273.

78. Raman, C. V., & Krishnan, K. S., (1928). The optical analogue of the Compton effect. *Nature, 121*(3053), 711.

79. Ravichandran, J., Yadav, A. K., Cheaito, R., Rossen, P. B., Soukiassian, A., Suresha, S. J., & Zurbuchen, M. A., (2014). Crossover from incoherent to coherent phonon scattering in epitaxial oxide superlattices. *Nature Materials, 13*(2), 168–172.

80. Regelman, D. V., Mizrahi, U., Gershoni, D., Ehrenfreund, E., Schoenfeld, W. V., & Petroff, P. M., (2001). Semiconductor quantum dot: A quantum light source of multicolor photons with tunable statistics. *Physical Review Letters, 87*(25), 257401.

81. Reiser, A., & Schächter, L., (2013). Geometric effects on blackbody radiation. *Physical Review A, 87*(3), 033801.

82. Ritz, T., Damjanović, A., & Schulten, K., (2002). The quantum physics of photosynthesis. *ChemPhysChem, 3*(3), 243–248.

83. Safronova, M. S., Kozlov, M. G., & Clark, C. W., (2011). Precision calculation of blackbody radiation shifts for optical frequency metrology. *Physical Review Letters, 107*(14), 143006.

84. Safronova, M. S., Porsev, S. G., Safronova, U. I., Kozlov, M. G., & Clark, C. W., (2013). Blackbody-radiation shift in the Sr optical atomic clock. *Physical Review A, 87*(1), 012509.

85. Salehi, H., & Bisabr, Y., (2003). *Conformal Invariance and Quantum Nature of Particles (No. hep-th/0301208).*

86. Schwartz, J. M., Stapp, H. P., & Beauregard, M., (2005). Quantum physics in neuroscience and psychology: A neurophysical model of mind-brain interaction. *Philosophical Transactions of the Royal Society B: Biological Sciences, 360*(1458), 1309–1327.

87. Shakhmuratov, R. N., Vagizov, F. G., & Gaiduk, V. Y., (2020). Methods of coherent control of spectral and temporal properties of gamma photons and their potential applications. *Crystallography Reports, 65*(3), 409–411.

88. Shpenkov, G. P., & Kreidik, L. G., (2004). Dynamic model of elementary particles and fundamental interactions. *Galilean Electrodynamics,* 23–29.

89. Simon, B., (1979). *Functional Integration and Quantum Physics* (Vol. 86). American Mathematical Soc.

90. Smoot, G. F., Gorenstein, M. V., & Muller, R. A., (1977). Detection of anisotropy in the cosmic blackbody radiation. *Physical Review Letters, 39*(14), 898.

91. Sorokin, A. A., Bobashev, S. V., Feigl, T., Tiedtke, K., Wabnitz, H., & Richter, M., (2007). Photoelectric effect at ultra-high intensities. *Physical Review Letters, 99*(21), 213002.

92. Spencer, W. P., Vaidyanathan, A. G., Kleppner, D., & Ducas, T. W., (1982). Photoionization by blackbody radiation. *Physical Review A, 26*(3), 1490.

93. Stapp, H. P., (1999). Attention, intention, and will in quantum physics. *Journal of Consciousness Studies, 6*(8, 9), 143–143.

94. Steinberg, R. N., Oberem, G. E., & McDermott, L. C., (1996). Development of a computer-based tutorial on the photoelectric effect. *American Journal of Physics, 64*(11), 1370–1379.

95. Strikman, M. I., & Frankfurt, L. L., (1980). Suppression of low-energy cascades as a result of the wave nature of slow particles. *Yadernaya Fizika, 32*(4), 968–971.
96. Sultanov, A. N., & Greenberg, Y. S., (2017). Effect of the qubit relaxation on transport properties of microwave photons. *Physics of the Solid State, 59*(11), 2103–2109.
97. Tanaka, H., Kawazoe, T., Ohtsu, M., Akahane, K., & Yamamoto, N., (2015). Evaluation of optical amplification properties using dressed photons in a silicon waveguide. *Applied Physics A, 121*(4), 1377–1381.
98. Tu, R. S., (2019). If the wave function collapse absolutely in the interaction, how can the weird nature of particles are born in the interaction?—A discussion on quantum entanglement experiments. *Indian Journal of Science and Technology, 12*, 8.
99. Vokos, S., Shaffer, P. S., Ambrose, B. S., & McDermott, L. C., (2000). Student understanding of the wave nature of matter: Diffraction and interference of particles. *American Journal of Physics, 68*(S1), S42–S51.
100. Walmsley, I., & Rabitz, H., (2003). Quantum physics under control. *Physics Today, 56*(8), 43–49.
101. Weymann, R., (1965). Diffusion approximation for a photon gas interacting with a plasma via the Compton effect. *The Physics of Fluids, 8*(11), 2112–2114.
102. Widger, Jr. W. K., & Woodall, M. P., (1976). Integration of the Planck blackbody radiation function. *Bulletin of the American Meteorological Society, 57*(10), 1217–1219.
103. Wilcox, R. M., (1967). Exponential operators and parameter differentiation in quantum physics. *Journal of Mathematical Physics, 8*(4), 962–982.
104. Yukawa, H., (1935). On the interaction of elementary particles. I. *Proceedings of the Physico-Mathematical Society of Japan. 3rd Series, 17*, 48–57.
105. Zeilinger, A., (1999). Experiment and the foundations of quantum physics. In: *More Things in Heaven and Earth* (pp. 482–498). Springer, New York, NY.

CHAPTER 4

ELEMENTARY PARTICLE PHYSICS

CONTENTS

4.1. INTRODUCTION

Humanity has long sought simplicity in the natural world. The ancient Greeks attempted to characterize the whole physical universe in terms of air, earth, fire, and water. To characterize the physical universe of time, matter, and space, these were substituted with the basic standards of mass, length, time, and charge. We saw that time and space are structure in the same foot, a representation of a single amount of time and space and that energy and matter are convertible, to the point where energy may be considered one of the basic quantities. We also discovered that energy is finite, thus matter must likewise be finite (Budker and Skrinskiĭ, 1978; Glashow, 1980). What is the lowest possible unit of matter? In other words, what are the basic or elemental components of matter?

What forces are at work on such basic particles? Is it feasible to unite such natural forces into a single integrated force that accounts for all observable interactions? In this chapter, we will try to address such questions (Gatti and Manfredi, 1986; Passon et al., 2018).

4.2. PARTICLES AND ANTIPARTICLES

Democritus and Leucippus, Greek philosophers, proposed that matter is made up of basic or essential particles known as atoms. The theory gained scientific support when John Dalton published A New System of Chemical Philosophy in 1808, wherein the enumerated around 20 chemical elements, every composed of one atom. There were about 60 known elements in 1896. It became clear that there had to be a mechanism to organize such distinct atoms to make sense of what had been swiftly devolving into chaos. Dimitri Mendeleev, a Russian scientist, produced the periodic table of the elements in 1869, depending on the chemical characteristics of the elements. The turmoil of the great variety of components was kept under control. The periodic table's unfilled spots have been used to anticipate new chemical elements. The atom might no longer be regarded as elementary when the interior structure of the atom was discovered (Okun, 1998; Plotnitsky, 2021).

The proton, electron, photon, and neutron were the only four essential particles recognized in 1932. Things seemed straightforward once more. This ease, however, was not to continue. In cosmic radiation, other particles had been quickly found. Cosmic radiations are cosmic particles that collide with the top of the atmosphere. Certainly, make it to the earth's surface, while others decompose into other particles before reaching the surface. Additional new particles had been discovered in man-made big accelerating devices. Hundreds of these particles exist now (Peskin, 2008; Kazakov, 2019).

Except for the neutron, proton, and electron, the majority of these fundamental particles disintegrate rapidly. We are back while attempting to bring order to the chaos of several particles. The 1st effort at order is the categorization of particles using the approach given in Figure 4.1, which groups all of the fundamental particles into leptons or hadrons (Szczekowski, 1989; Nagashima, 2011).

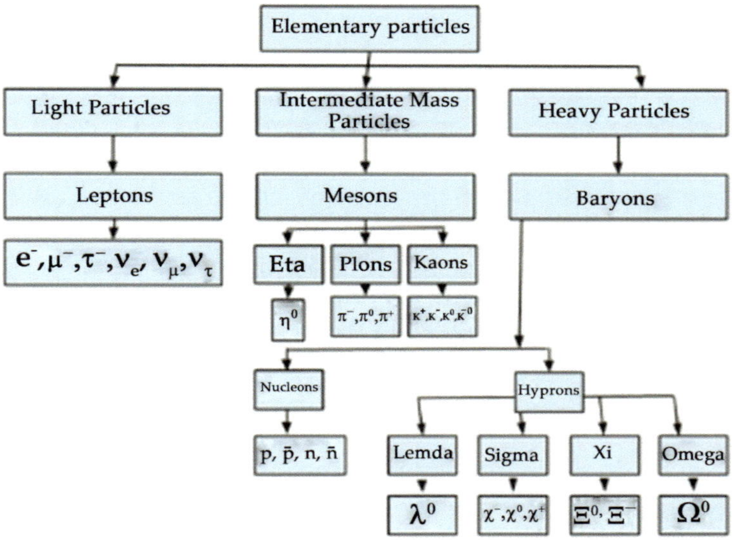

Figure 4.1. The first categorization of the essential particles.

Source: https://kullabs.com/class-12/physics/particle-physics-and-cosmology/introduction-to-elementary-particles.

4.2.1. Leptons

The Leptons are particles that are not affected by the strong nuclear force. They are relatively tiny in size, around 10^{-19} meters in diameter. They are all have spin ±½, in units of $\hbar$. There have been 6 leptons in total: the muon, μ^-, the electron, e^-, and the tauon, τ^-, with its neutrino. They may take any shape (Omnès, 1972; Wesson, 2013).

$(v_e)\ (v_\mu)(v_\tau)$

$(e^-)\ (\mu^-)\ (\tau^-)$ (1)

Therefore, there have been three neutrinos: the muon-linked neutrino, v_μ; the electron–neutrino, v_e; and the tauon-linked neutrino v_τ. The muon is extremely similar to the electron but so much heavier. It has a mass approximately 200 times that of an electron. This is not steady like the electron and decay in around 10^{-6} seconds.

Initially, the term lepton, which derives from the Greek term *leptos* meaning little or lightweight, denoted the lightness of such particles. Furthermore, the τ lepton had been found in 1975, and its mass is double that of the proton. In other words, the τ lepton is massive, and a misnomer (Zel'dovich and Khlopov, 1981; Skrinskiĭ, 1982).

Leptons are basic in the sense that they lack of structure. Such that, they do not consist of anything smaller. Leptons interact under the weak nuclear force, whereas the charged leptons, e^-, μ^-, τ^-, also contribute to the electromagnetism interaction (Kopylov and Podgoretskij, 1975; Drell, 1977).

Initially, the muon was believed to be Yukawa's meson, which mediated the strong nuclear force; thus, it was referred to as a μ^- meson. It is now recognized to be incorrect, as the muon is a lepton and not a meson.

4.2.2. Hadrons

Hadrons are subatomic particles influenced by the powerful nuclear force. There are a large number of specific hadrons. Hadrons contain an interior structure made of quarks, which seem to be elementary particles. Mesons and baryons are two subgroups of hadrons that may be further subdivided.

- **Baryons:** These are heavy subatomic particles whose decay products are composed of at least one proton or neutron. The spin of baryons is half-integral, such that, 1/2 $\hbar$, 3/2 $\hbar$, etc. In a moment, it would become clear that all baryons are made of three quarks (Bernstein and Feinberg, 1962; Zichichi, 1979).
- **Mesons:** These had been initially intermediate-mass particles between the proton and the electron. Furthermore, several enormous mesons have subsequently been discovered, rendering the original description obsolete. Mesons are currently defined as particles whose disintegrated products do not contain baryons (Recami and Rodrigues, 1985). We shall see that mesons are

formed of a pair of quarks and antiquarks. All mesons possess intrinsic spin, or 0, 1, 2, 3, etc. The mass of a meson grows as its spin rises. In Table 4.1, several fundamental particles are listed.

Table 4.1. Some of the Basics Fundamental Particles

Leptons	muon electron neutrinos tauon	μ^- e^- ν_e, ν_μ, ν_τ τ^-
Hadrons		–
Mesons	pi rho eta omega delta phi	π ρ η Ω Δ ϕ
Baryons	neutron proton lambda delta Hyperon Sigma Omega	n p λ Δ Λ Σ Ω

Paul Dirac used special relativity and quantum theory to create a relativistic theory of the electron in 1928. The calculations anticipated two energy levels for every electron, which was a surprise outcome of the merger. One is linked to the electron, while the other is linked to a particle that is identical to the electron apart from that it has a +ve charge. The antielectron, or positron, was the name given to this article. It was the earliest forecast of antimatter's presence. In 1932, the positron was discovered (Castell, 1966; Kopylov and Podgoretsky, 1975).

There must be an antiparticle for each particle found in nature. The antiproton is the proton's antiparticle. It possesses all of the properties of a proton except for the fact that this is negatively charged. The π^0 mesons and the photon are examples of totally neutral particles that have their antiparticles. Antiparticles are represented by a bar over the particle sign. As a result, p, and n are antiprotons and antineutrons, respectively. Antimatter is made up of antineutrons, antiprotons, and antielectrons (positrons) while matter is made up of neutrons, protons, and electrons. Figure 4.2 depicts antimatter and matter atoms. Antimatter is kept together by similar electric forces which keep matter together (Chiu, 1966; Toussaint and Wilczek, 1983).

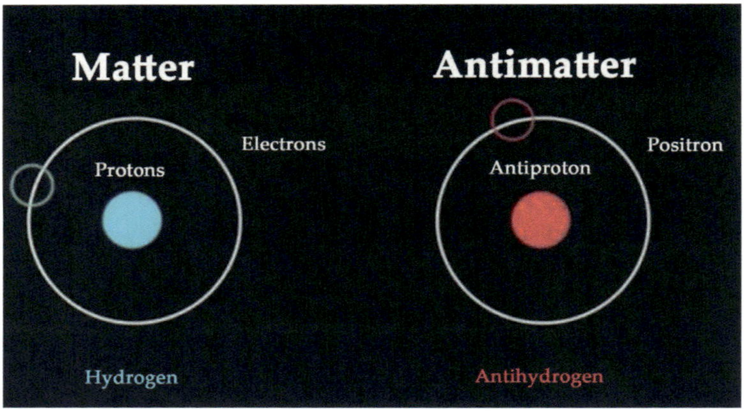

Figure 4.2. Antimatter and matter.

Source: https://www.indiatoday.in/education-today/gk-current-affairs/story/where-is-all-the-antimatter-scientists-create-antihydrogen-atom-to-find-answers-html-1206368-2018-04-06.

(Remember that in antimatter, the +ve and –ve signs are reversed). High-energy accelerators have previously produced the antihelium nucleus. When antiparticles and particles collide, they destroy one another, leaving just energy. When an electron collides with a positron, for instance, they annihilate as per the reaction (Smarandache, 2009; Adamczyk et al., 2013).

$$e^- + e^+ \rightarrow 2\gamma \qquad (2)$$

Photons of electromagnetic energy have been represented by 2γ's. (For energy and momentum conservation, two gamma rays are required). Other particles may be created with this energy. Particles, on either hand, may be formed by transforming the energy in a photon into a particle-antiparticle pair, like:

$$\gamma \rightarrow e^- + e^+ \qquad (3)$$

A *Feynman diagram*, named after the American physicist Richard Feynman (1918–1988), may depict destruction or formation, as depicted in Figure 4.3. The formation of an electron-positron pair is seen in Figure 4.3(a).

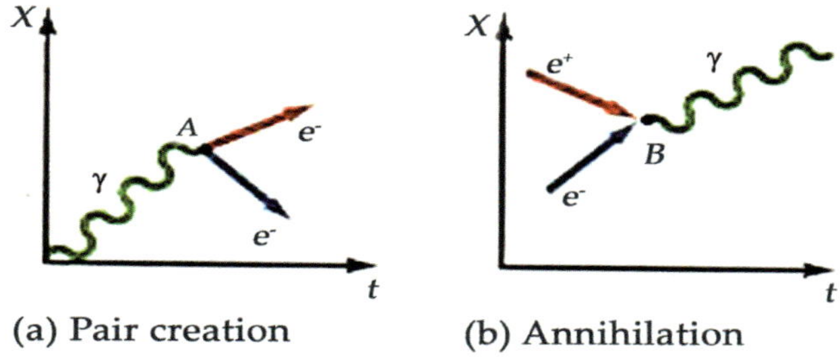

Figure 4.3. Particle creation and annihilation.

Source: https://energywavetheory.com/explanations/particle-annihilation/.

The photon travels across spacetime till it approaches spacetime point A when its energy is transformed into an electron-positron pair. Figure 4.3(b) depicts a positron and electron interacting at spacetime point B, wherein they destroy one another and only the photon flows via spacetime. (To conserve energy and momentum throughout the formation procedure, a quite massive nucleus is necessary) (Knudsen and Reading, 1992; Martín-Martínez and Fuentes, 2011).

4.3. THE FOUR FORCES OF NATURE

Four forces that operate on matter particles have been recognized in natural science.

They are as follows:

- **The First is the Gravitational Force:** The gravitational force is the earliest force that has been discovered. It binds us to the earth's surface and the whole cosmos together. This is an along-range force with a variation of $1/r^2$. When contrasted to the other natural forces, this is by far the smallest.
- **The Electromagnetic Force is Number Two:** The electromagnetic force had been discovered second. Actuality, before the initial unification of the forces, was two forces: magnetic and electric. Molecules, atoms, liquids, and solids are held together by electromagnetic force. It is a long-range force that varies as $1/r^2$ like gravity.
- **The Weak Nuclear Force is Number Three:** The weak nuclear force reveals itself not so much in keeping stuff together as it does

in enabling it to dissolve, as seen in proton and neutron decay. By allowing a proton to decay into a neutron as shown in Eqn. (3), the weak force is accountable for the process of fusion that occurs in the sun. The proton-proton cycle will continue till helium is created, at which point massive amounts of energy have been released. The nucleosynthesis of the chemical element was also triggered by the weak force. The weak nuclear force, unlike the electromagnetic and gravitational forces, has a relatively small range (Bruschi et al., 2012; Eseev and Meshkov, 2016).

- **The Strong Nuclear Force is Number Four:** The strong nuclear force is accountable for the nucleus's stability. It is the most powerful of all the forces, yet it has a relatively limited range. Such that, its impact is felt within a radius of around 10–15 m, the nucleus' diameter. There is no indication of its presence at distances larger than this. Only hadrons are affected by the strong nuclear force.

After combining time and space into spacetime, Einstein attempted to combine the electromagnetic and gravitational forces into a single force. Despite spending a lifetime attempting, he was unsuccessful. However, the prospect of bringing the armies together has not gone. Indeed, as we will see soon, Salam, Weinberg, and Glashow theoretically united the weak nuclear forces and electromagnetic forces into the electroweak force, which Rubbia empirically validated. A grand unification of the strong and weak electric forces is suggested. Ultimately, an effort is being made to unite these four forces into a single superpower (Meiman, 1964; Brandenburg et al., 1975).

4.4. QUARKS

George Zweig and Murray Gell-Mann separately suggested in 1964 that hadrons were not elementary particles, but instead were formed of even more elementary particles, to bring order to the vast number of elementary particles. Such particles were named quarks by Gell-Mann. He thought there were just three of them at first, but that number has now grown to 6. Table 4.2 depicts the 6 quarks. Down, up, charmed, strange, top, and bottom are the names of the quarks. Such quarks carry fractional electric charges, which is one of their features (Braun, 2001; Araki et al., 2021).

Table 4.2. The Quarks

Name (Flavor)	Charge	Spin	Symbol
Charmed	2/3	½	C
Strange	–1/3	½	S
Top	2/3	½	T
Bottom	–1/3	½	B
Down	–1/3	½	D
Up	2/3	½	U

Such that, the bottom, strange, and down quarks each have 1/3 of the charge of an electron, while the top, up, and charmed quarks each have 2/3 of the charge of a proton. They are all spin 1/2 in units of. Every quark has an antiquark, that is identical to the original but has the opposite polarity. The antiquark is represented as a bar above the letter q (Bourquin et al., 1979; Abdel-Raouf, 1987).

We will see now that quarks make up all of the hadrons.

Three quarks make up baryons:

Baryon = qqq (4)

The mesons, on the other hand, are composed of a quark-antiquark pair:

Meson = qq (5)

Consider the proton as an instance of a baryon formed from quarks. As seen in Figure 4.4(a), the proton is made up of one down quark and two up quarks (Azimov and Ioganesen, 1981; Abov et al., 1984).

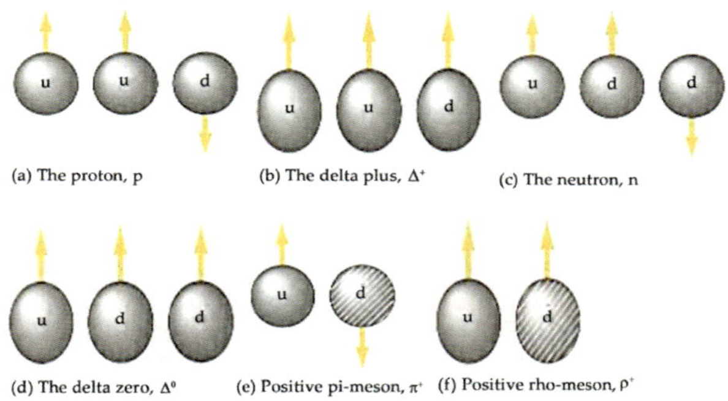

Figure 4.4. Certain meson and baryon quark configurations.

Source: https://courses.lumenlearning.com/physics/chapter/33-5-quarks-is-that-all-there-is/.

Adding the charges of the constituent quarks yields the proton's electric charge. This is because the d quark contains a charge of –1/3 and u quark contain a charge of 2/3, the proton's charge is, as anticipated (Aloisio et al., 2004; Cremaschini and Tessarotto, 2013).

$2/3 + 2/3 - 1/3 = 1$

The proton's spin must now be 1/2 in units of. The two up quarks in Figure 4.4(a) have their spins raised due to the orientation of the arrow upon the quark. The arrow on the down quark points down, indicating that its spin is down. Since every quark has 1/2 spin, the proton's spin is calculated by combining the spins of the quarks as:

$1/2 + 1/2 - 1/2 = 1/2$

Notably, the names down and up for quarks have nothing to do with their spin orientation. As seen in Figure 4.4, the delta plus +baryon is composed of the same three quarks as the proton, but its spins have all been oriented in a similar direction (b). Therefore, the Δ^+ particle's spin is $1/2 + 1/2 + 1/2 = 3/2$.

That means the Δ^+ particle's spin is 3/2. Because aligning the spins in a similar direction consumes more energy, quark spins that are aligned have greater energy. Einstein's equivalence of mass-energy ($E = mc^2$) displays this as an increment in mass. As a result, the Δ^+ particle's mass is greater than the proton. As a result, to create particles from quarks, we must not only recognize the sorts of quarks that make up the particle but also the direction of its spin (Barnett and Loudon, 2006; Chaichian et al., 2012).

A neutron is composed of 2 down quarks and 1 up quark, as shown in Figure 4.4(c). The entire charge of electricity is:

$2/3 - 1/3 - 1/3 = 0$

Although it has a unique spin,

$1/2 + ½ - 1/2 = 1/2$

Figure 4.4(d) demonstrates that the delta zero Δ^0 particles are composed of similar three quarks, however, its spins have all been oriented. Consider the pi plus π^+ meson in Figure 4.4 as an instance of the production of a meson from quarks (e). It is composed of an anti-down quark and an up quark. Its charge is expressed as:

$2/3 + [-(-1/3)] = 2/3 + 1/3 = 1$

Thus, the d quark contains a charge of –1/3, whereas its antiquark d contains a similar charge however the opposite sign, The spin of the π^+ is

1/2–1/2 = 0.

Therefore, the π^+ meson contains a charge of +1 and zero spins. If the spins of such two quarks have been oriented, as shown in Figure 4.4(f), the meson is the +ve rho-meson ρ+ having a spin of 1 and an electric charge of +1 (Takatsu et al., 1988; Mansuripur et al., 2013).

Certain baryons' quark structures correspond to the quark structures of certain mesons. Strange particles are particles that possess a strange quark. The term comes from the fact that such particles decayed very much slower than the other fundamental particles that it had been thought strange. We would want to "see" the quarks that make up a neutron or proton. We may "see" within a proton via bombarding it with neutrinos or electrons, much as Rutherford "see" within the atom via hitting it with alpha particles. Protons had been bombarded via high-energy electrons at the Stanford Linear Accelerator Center (SLAC) in 1969 (Tomura and Kunieda, 2009; Ahn et al., 2011). A few of these electrons had been discovered to be dispersed at quite large angles, similar to Rutherford scattering, suggesting that the proton has tiny components. Figure 4.5 depicts a proton's image as seen by scattering tests. Particles having charges of –1/3 or +2/3 of the electrical charge seem to be scattering. (Remember that the down quark is charged –1/3, while the up quark is charged +2/3). As a result, there has been empirical evidence supporting the proton's quark structure. Similarly, successful tests are carried out on neutrons. The scattering also indicated that the proton contains certain quark-antiquark couples. Remember that mesons are made up of quark-antiquark pairs. The experiments also revealed the presence of gluons, which are particles that reside inside nucleons. The gluons are quark-to-quark exchange particles that act to keep the quarks united. They are the nuclear adhesive (Kumar and DebRoy, 2003; Qiu et al., 2020).

The only problem with the quark model right now is that it appears to violate the principle of Pauli exclusion. Remember that no two electrons may have a similar quantum number simultaneously, as per the principle of Pauli exclusion. The principle of Pauli exclusion is more comprehensive than that, since it applies to any particle with half-integral spins, like 1/2, 3/2, 5/2, etc. Fermions are particles that contain half-integral spin. Quarks should also follow the principle of Pauli exclusion since their spin is 1/2. The Δ++ particle, on the other hand, is made up of three up quarks that contain a similar spin, while the Ω-particle is made up of three strange quarks that contain a similar spin (Yoshikawa and Morita, 2005; Mansuripur, 2008). As a result, every quark should have an extra feature that is unique to it to avoid violating the principle of Pauli exclusion. The quark's new property is named "color."

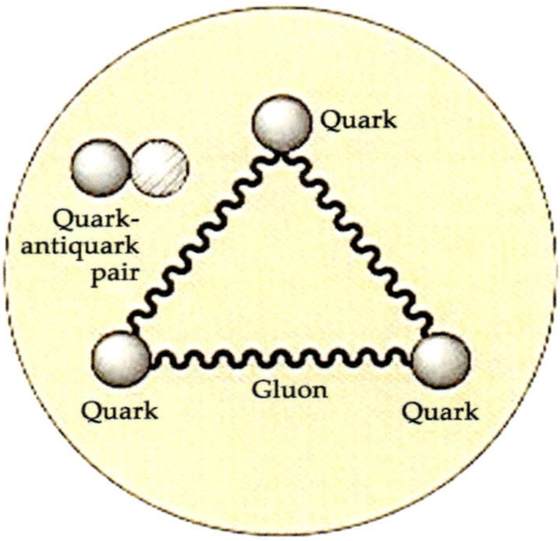

Figure 4.5. The proton's structure.

Note: Inside the proton by D. H. Perkins.

Source: https://www.researchgate.net/publication/231034177_The_quark_model.

Red, blue, and green are the three hues of quarks. It is important to remember that such colors are merely labels and have nothing to do with the actual colors we see daily with our eyes. The words have been chosen at random (Maxwell, 1868; Ahn et al., 2012). They may have been referred to as A, B, and C, for instance. Color may be thought of similarly to electric charges. There are two types of electric charges: +ve and –ve charge. Color charges are available in three colors: blue, green, and red. The blue-up quark uB, the green-up quark uG, and the red-up quark uR are the three varieties of up quarks. As a result, the delta plus-plus particle Δ++ may be depicted in Figure 4.6(a). Because every up quark is unique, there is no violation of the principle of Pauli exclusion (Kim and Lee, 2000).

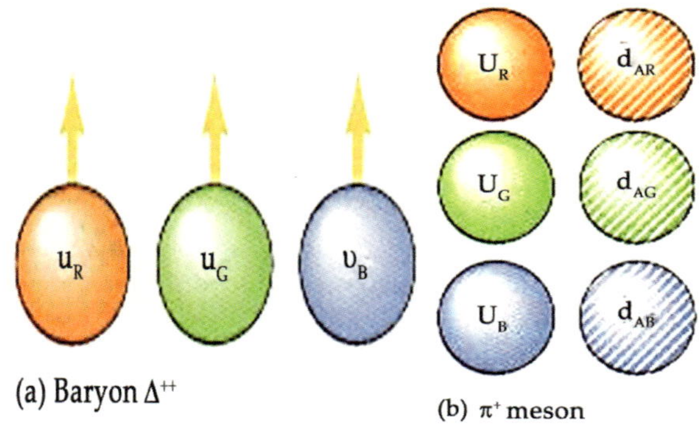

Figure 4.6. Colored quarks.

Source: https://www.forbes.com/sites/startswithabang/2019/04/18/quarks-dont-actually-have-colors/.

Green, blue, and red quarks make up all baryons. The combination of green, blue, and red quark is supposed to form the color white, even as the main colors green, blue, and red add up to white. As a result, all baryons are described as colorless or white. Every color of quark has an anti-color, just like a quark contains an antiquark. As a result, an antired-up quark is a red-up quark with an up antiquark that bears the color antired. Flavors are several types of quarks, like strange, down, up, and so on. As a result, every quark flavor is available in three colors, for a total of 6 tastes multiplied by three colors equals 18 quarks. There are 18 antiquarks linked with the 18 quarks. Mesons, like baryons, should be white or colorless (Arkkio et al., 2000; Lin et al., 2016).

Because a color plus its anti-color equals white, one colored quark of a meson ought to be linked with an anti-color. As a result, Figure 4.6 depicts various π^+ meson structures (b). To generate the white π^+ meson, a red-up quark uR joins with an anti-down quark carrying the color antired dAR. (In Figure 4.4, the anticolor quark is represented by hatching lines.) Likewise, the uRdAR+ uGdAG+ uBdAB, π^+ meson may be produced from green and antigreen uGdAG and blue and antiblue uBdAB quarks, as well as linear combinations of them. Eqns. (4) and (5) may be rewritten as:

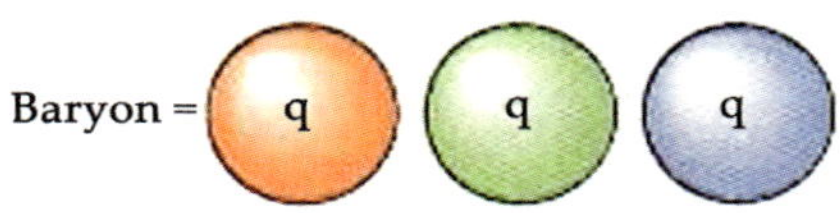

(6)

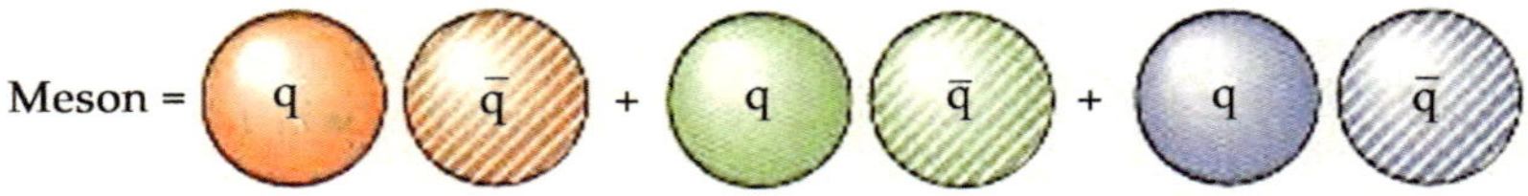

(7)

The attraction between a color-carrying quark and its antiquark-carrying anticolor antiquark has always been strong. Similarly, the attraction among three quarks of various colors is appealing. All other color combinations have a repellent effect. When we talk about the strong nuclear force, we will talk more about colored quarks (Zakharian et al., 2005; Xiong et al., 2019).

4.5. THE ELECTROMAGNETIC FORCE

To recap the findings, Coulomb's law defined the electric force among charged particles, and the electric field served as the force's mediator. Ampère established the connection between electric and magnetic fields when he realized that a current running through a wire created a magnetic field. A shifting magnetic field created an electric current, according to Faraday. In his renowned electromagnetism equations, James Clerk Maxwell combined all of electricity and magnetism. In other words, the distinct forces of electric and magnetic fields were combined into a single electromagnetic force (Engel and Friedrichs, 2002; Qiu et al., 2019).

The combination of electromagnetic theory and quantum mechanics has resulted in quantum electrodynamics (QED). The electric force is conveyed in QED via virtual photon exchange. As seen in Figure 4.7, the force between two electrons may be viewed. Remember that the Heisenberg uncertainty relation permits the production of a virtual particle if the energy linked with the virtual particle's mass is repaid in a period t that fulfills Eqn. (7). Two electrons approach each other in Figure 4.7. The initial electron recoils after emitting a virtual photon (Chaumet and Nieto-Vesperinas, 2000; Berger et al., 2019).

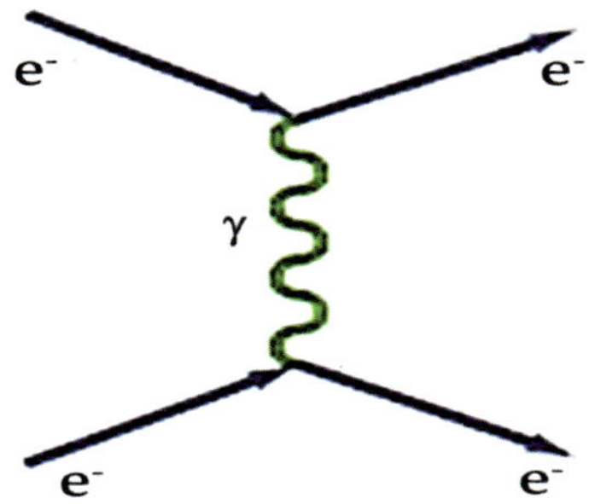

Figure 4.7. The electric force as an exchange of a virtual photon.

Source: http://abyss.uoregon.edu/~js/21st_century_science/lectures/lec17.html.

When the second electron absorbs the photon, it likewise recoils, as indicated, resulting in a repulsive force between the two electrons as a consequence of the photon interchange. This exchange force is simply a quantum mechanical phenomenon with no true classical parallel. As a result, visualizing how a photon exchange between an electron and a proton causes an attractive attraction between them may be a bit more challenging. The force's mediator or transmitter is the photon that has been exchanged. A single traded particle may symbolize all of nature's forces (Salam, 1979; Wang et al., 2000).

Since the rest mass of a photon is zero, the range of the electric force is unlimited. This may be demonstrated using a few equations. The payback period for the uncertainty principle was:

$$\Delta t = \frac{\hbar}{E}$$

$$\Delta E = (\Delta \mathrm{m})c^2$$

Whereas the virtual particle's energy ΔE was linked to its mass Δm by:

$$\Delta E = (\Delta m)c^2$$

By putting this into the equation, the payback time was calculated:

$$\Delta t = \frac{\hbar}{(m)c^2}$$

The maximum distance a virtual particle might go and yet return in time Δt was calculated as:

$$d = c\,\frac{t}{(m)c^2}$$

The range of the virtual particle is the measurement of this distance. When you put the upper equation into Eqn. (7), you get the range.

$$d = c\,\frac{t}{2}$$

$$d = \frac{\hbar}{2c}\frac{1}{m} \qquad (8)$$

The rest mass m of a photon is equivalent to 0. As a fraction's denominator approaches 0, the fraction approaches infinity. As a result, the particle's range d is infinite. As a result, the electric force must reach infinity, which it does (Portelli, 1987; Pangilinan, 2010).

4.5.1. The Weak Nuclear Force

The role of the weak nuclear force in radioactive decay is well-known. A neutron in the nucleus decays according to the relationship as the first stage of β-decay.

$n \rightarrow p + e^{-} + ve$ (9)

While the proton within the nucleus decays as:

$p \rightarrow n + e^{+} + ve$ (10)

The β+ decay begins with this phase. Ultimately, the reaction starts the radioactive degradation that is triggered by the nucleus capturing an electron (electron capturing).

$e^{-} + p \rightarrow n + ve$ (11)

These three reactions are only a few of the ones that the weak nuclear force mediates. The weak nuclear force is not a standard pull or push force as defined by classical physics. Instead, it is in charge of the subatomic particle transformation. The weak nuclear force operates among hadrons and leptons, as well as between hadrons and hadrons, and is irrespective of electric charge. The feeble nuclear force has a range of just around 10^{-17} meters. The decay duration is significantly longer, with the mild decay taking around 10^{-10} seconds and the strong interaction taking about 10^{-23} seconds (Kriske, 2011). After gravity, the weak nuclear force is the weakest. The neutrino is a result of weak interactions. Neutrinos are extremely small particles. Some believe they have no rest mass, whereas others believe they are very tiny, with a neutrino maximum limit of roughly 10^{-30} eV. Only the weak force affects the neutrino, neither the stronger nor the electromagnetic forces. It may pass via earth or the sun without ever engaging with anything since its engagement is so weaker (Lee, 1991; Porsev et al., 2009).

4.6. THE ELECTROWEAK FORCE

Sheldon Glashow, Abdus Salam, and Steven Weinberg earned the 1979 Nobel Prize for their proposal to unify the electromagnetic force having the weak nuclear force. Such force is referred to as the electroweak force. Just like a virtual photon mediator the electromagnetic force among charged particles, it became evident that certain particles must likewise mediate the weak nuclear force.

The novel electroweak force is transmitted via four particles: the photon and the vector bosons W^+, W^-, and Z^0. The photon mediators the electromagnetic force, while vector bosons transmit the weak nuclear force. Figure 4.8 depicts the neutron decay Eqn. (9) in the context of the exchange particles (a). A neutron decays by releasing a W-particle, therefore transforming into a proton. Within 10^{-26} s, the W-particle decays into an antineutrino and an electron. Figure 4.8 depicts the decay of the proton in a radioactive nucleus, as described by Eqn. (11). The proton releases the intermediary vector boson W^+ and transforms into a neutron. The W^- eventually decays into a neutrino and a positron. Figure 4.8(c) depicts an electron capture, defined by Eqn. (11), as an interaction between an electron and a proton. The proton releases a W^+ and is transformed into a neutron. The W^+ then interacts with the electron to generate a neutrino. As seen in Figure 4.8(d), the Z^0 particle is detected in electron-neutrino scattering (Kieu, 1994; Fredsted, 2010).

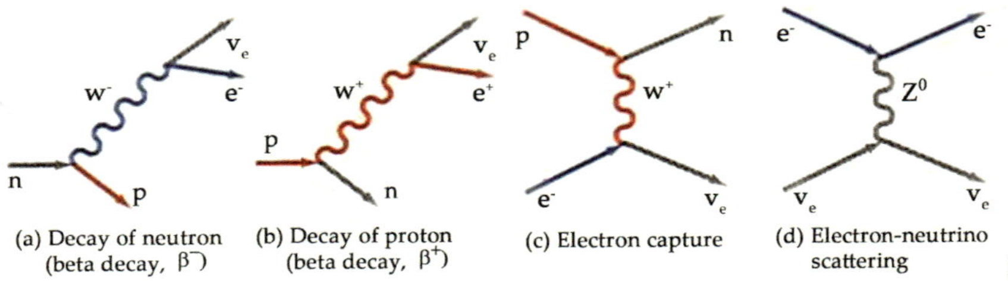

Figure 4.8. Examples of the electroweak force.

Source: https://www.nuclear-power.com/nuclear-power/reactor-physics/atomic-nuclear-physics/fundamental-interactions-fundamental-forces/electroweak-interaction/.

In January 1983, a team led by Carlo Rubbia of Harvard University discovered the vector bosons W^- and W^+ in proton-antiproton collisions at a higher energy at the European Center for Nuclear Research (CERN). The Z^0 was discovered a few months later, in May 1983. The W had a mass of roughly 80 GeV, whereas the Z^0 had a mass of around 90 GeV. We can see from Eqn. (8) that for such a huge mass, Δm in that equation offers a relatively small range for the weak force, as empirically discovered (Thomas, 2006; Fredsted, 2007).

The electromagnetic and weak nuclear forces combine at very higher energies, approximately 100 GeV, to form one electroweak force that operates equally on all particles: leptons and hadrons, uncharged, and charged (Mason, 1991; Gleeson and Morley, 1992).

4.6.1. The Strong Nuclear Force

The strong nuclear force is accountable for binding neutrons and protons together in the nucleus, as previously stated. To counteract the huge electrical force of repulsion between the protons, the strong nuclear force should be extremely strong. The nuclear force, according to Yukawa, is created by the interchange of mesons among nucleons. The nucleons, on either hand, are composed of quarks (Aaltonen et al., 2018).

What is it that binds these quarks together?

The electric force in QED is created by the interchange of virtual photons. The force binding quarks together is created by the exchange of a new particle dubbed a "gluon," according to quantum chromodynamics (QCD), one of the most recent theories in fundamental particle physics. A gluon is a nuclear glue that keeps quarks in a nucleon together. Figure 4.9(a) depicts the force among quarks as a virtual gluon exchange. Like quarks, gluons have colors and anticolors. When a gluon interacts with a quark, the color of the quark changes. Figure 4.9(b) depicts a red-up quark uR producing a red-antiblue gluon RB as an instance. The color of the up quark changes from red to blue. That is, even if an anti-color is removed, the color should remain. As a result, even if an antiblue is removed from the up quark, the color blue should remain (Luo et al., 2009; Bock, 2019). The blue of the up quark interacts with the antiblue of the gluon to cancel out the color blue whenever it gets the red-antiblue gluon RB. (White is always the result of color and its anticolor).

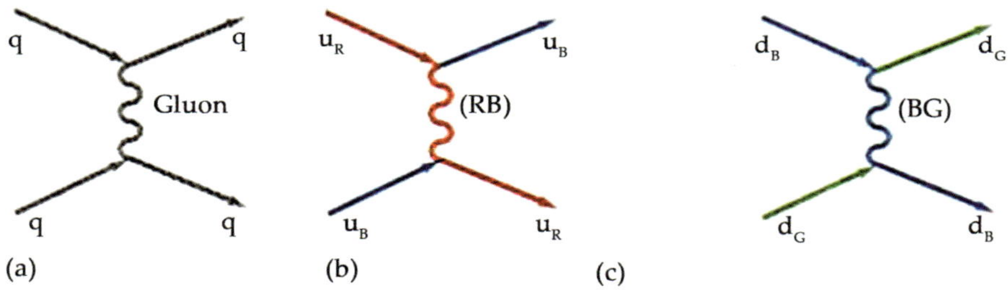

Figure 4.9. Gluons are exchanged between quarks.

Source: http://hyperphysics.phy-astr.gsu.edu/hbase/Particles/expar.html.

The up quark absorbs the gluon's red hue, transforming it into a red-up quark. The quarks changed hue as a result of transferring the gluon. Figure 4.9(c) depicts a blue-down quark producing a blue-antigreen gluon BG, converting it to a green-down quark. The color green cancels when the initial green-down quark absorbs the BG gluon, and the down quark becomes a blue-down quark (Reifler and Morris, 1985; Carlip, 2008).

There are eight distinct gluons in all, each with its mass. One color and one anticolor are always carried by every gluon. A gluon may occasionally convert into a quark-antiquark pair.

Scattering from protons provides much more information at energies higher than those utilized in the scattering illustrated in Figure 4.9, as seen in Figure 4.10. The three valence quarks are still visible, but there are a lot more quark-antiquark couples this time (Boulware and Deser, 1975; Goldman et al., 1986).

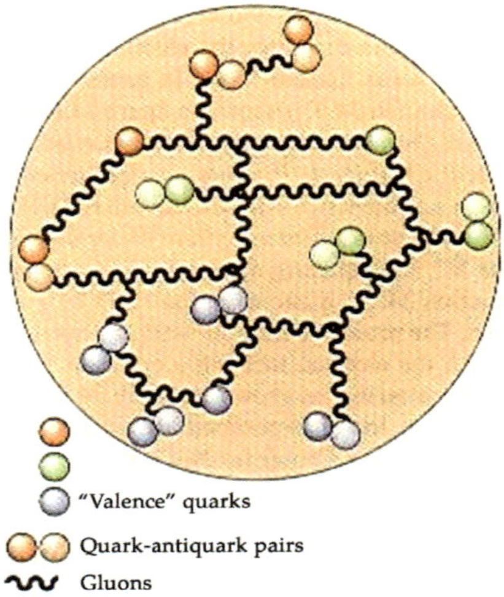

Figure 4.10. Proton structure in greater depth.

Note: Adapted from: D. H. Perkins.

Source: https://onlinelibrary.wiley.com/doi/10.1002/tea.20324.

Remember that a meson is made up of a quark-antiquark pair. As a result, the proton is filled with imaginary mesons. The gluons are also visible. To address the conventional issue of what binds protons together in the nucleus, the strong force is the product of color forces among quarks inside the nucleons. The quark-antiquark pair (meson) generated by the gluons are transferred between the nucleons at relatively significant spacing distances inside the nucleus. The strong force may be interpreted at small distances inside the nucleus as an interchange of quarks from one proton containing a quark from another proton, or as a specific exchange of the gluons themselves, that give birth to the quark-quark interaction inside the nucleon. The strong force is therefore created by quarks, and the force that holds neutrons and protons together in the nucleus is a manifestation of the interaction among quarks (Mattingly, 2005; Smolin, 2010).

Why have quarks never been segregated unless they are the building blocks of all hadrons? The quark-quark force is similar to the elastic force $F = kx$ described by Hooke'slaw. The force between the quarks is negligible for small values of the separation distance x, and the quarks are largely free to move about within the particle. When we disassociate the quarks across a vast gap distance of x, although, the force becomes extremely high, to the point where the quarks may not be detached at all. The confinement of quarks is the name for this situation. Quarks may not depart from the particle in which they are components, hence they are never observed in isolation (Isham, 1995; De Aquino, 2002).

Is there, however, any proof that quarks exist? Yes, it is correct. In 1978, experiments were carried out at DESY (Deutsches Elektronen-Synchrotron) in Hamburg, Germany, in the modern PETRA storing ring. In a head-on collision, positrons, and electrons with energies of 20 GeV each were shot at each other. The disintegration of the electron and its antiparticle, the positron, releases a tremendous amount of energy, which is used to create quarks. Exactly as expected, the experimenters discovered a sequence of "quark jets," which were the decay byproducts of the quarks. (A quark jet is a group of hadrons that leave the interaction in a similar direction.) The presence of quarks was indirectly demonstrated by such quark jets. CERN as well as other accelerators have conducted similar tests (Gorelik, 1992; Rovelli, 2003).

Gluons and quarks do not appear to be built up of smaller particles once they had been established at this time; they seem to be elementary. Certain speculative hypotheses, although, claim that quarks are composed of even tiny particles known as "preons." Moreover, there is no proof of the presence of preons at this time (Albers et al., 2008; Calcagni, 2020).

4.6.2. Grand Unified Theories (GUT)

If the electric and weak nuclear forces can be combined to form a single electroweak force, then why doesn't the strong nuclear force? Howard Georgi and Sheldon Glashow achieved just that in 1973 when they published a hypothesis that combined electroweak and strong forces. The grand unified theory (GUT), was given to this new idea as the first of several (Hooft, 1993; Gorelik, 2005).

The weak nuclear force was linked to the strong nuclear force in the first stage of this merger. Consider the neutron's decay, as given in Eqn. (9):

$$n \rightarrow p + e^- + ve$$

According to the illustration in Figure 4.11, we may now view this degradation. The quark hypothesis states that a neutron consists of two down quarks and 1 up quark.

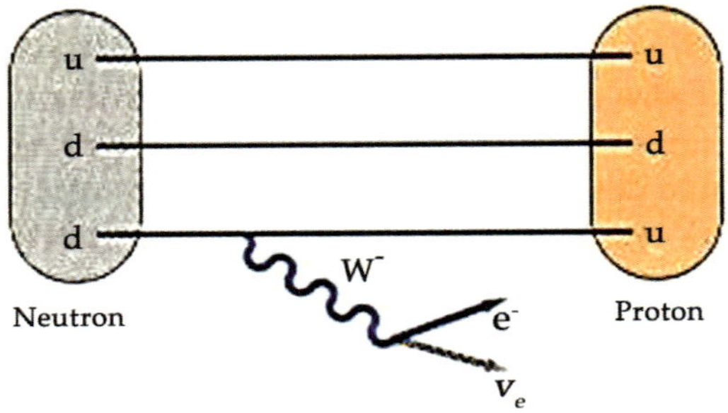

Figure 4.11. The decay of the neutron.

Source: http://hyperphysics.phy-astr.gsu.edu/hbase/Particles/proton.html.

The W^- boson is emitted by one of the neutron's down quarks, which causes the neutron to be transformed into a proton. (Remember that the proton is made up of two up quarks and one down quark.) Following that, the W^- boson decays into an antineutrino and an electron. As a result, the weak force alters a quark's taste while the strong force alters merely its color (Basov et al., 1961; Childs et al., 1986).

We may no longer distinguish between the weak, strong, and electromagnetic forces above 10^{15} GeV of energy, known as the *grand unification energy*.

There is just one coherent interaction or force that happens above this energy. Of course, this energy is far bigger than anything we might ever hope to achieve through experimentation. Although, it may have been achieved during the early phases of the universe's birth, the so-called "Big Bang."

The strong nuclear force acts on quarks, while the weak nuclear force acts on leptons and quarks. Leptons and quarks must be facets of one more basic quantity if the weak and strong forces are to be merged. Such that, the grand unified force must have the ability to convert quarks into leptons and vice versa. The united force is mediated by 24 particles in the GUT, which are mentioned in Table 4.2. Forces are united in the GUT because they result from the interchange of the same family of particles. The photon transmits the electromagnetic force; the vector bosons transmit the weak force; the gluons transmit the strong force, and there are now 12 new particles known as X particles (also known as progenitor and/or lepto-quark particles) that mediate the unified force. These X particles are able to turn quarks into leptons, transforming hadrons into leptons (Coldren et al., 2004; Coleman, 2012).

The electrical charges of the X particles range from 1/3 and 4/3. As a result, the X particles are $X^{1/3}$, $X^{-1/3}$, $X^{4/3}$, and $X^{-4/3}$. Such X particles are also available in the three colors green, blue, and red for a total of 12 X particles. Figure 4.11 shows how the X particles may convert a quark into a lepton. An X particle with a color charge of antired and an electrical charge of –4/3 unites with a red-down quark with an electrical charge of 1/3. Whereas the electrical charge becomes 1/3 – 4/3 = –3/3 = –1 and an electron are generated out of a quark, the colors red and antired cancel to obtain white. Since the mass of the virtual X particle should be on the order of 10^{15} GeV, which is an incredibly big energy, this sort of process is not easily seen in ordinary life. A single proton must similarly decay, according to a similar analysis. However, the expected lifespan is 10^{32} years. Experiments are being carried out to see if the projected decays occur. Furthermore, no such decay of an individual proton has yet been discovered. An isolated proton appears to be a relatively stable particle, implying that additional research is needed or that the GUT model has to be modified (Daneu et al., 1986; Crowley et al., 2012).

4.7. THE GRAVITATIONAL FORCE AND QUANTUM GRAVITY

Newton discovered that cosmic and terrestrial gravity were the same and combined them into his universal gravitation law. It had been discovered, although, that it was not that universal. The shift began with Einstein's special theory of relativity, which limit systems traveling at constant velocity concerning each other. He discovered the equivalency among accelerated systems and gravity when he expanded this theory to systems that had been accelerated concerning each other. The next step was to demonstrate that matter distorted spacetime and that gravity was a representation of that warped spacetime. As a result, general relativity became a rule of gravity, and Newton's law of gravitation was shown to be nothing more than a specific instance of Einstein's theory of general relativity (De Loach et al., 1973).

We also learned that quantum theory is one of the most important new ideas in modern physics, implying that nature is quantized. Quanta of energy, angular momentum, mass, charge, and other things exist. However, in its current form, general relativity is largely independent of quantum theory. In this way, it is still classical physics. It, too, should be an approximation of nature's underlying reality. A broader theory would combine quantum mechanics and general relativity, resulting in a quantum gravity theory.

To link quantum theory with general relativity (also known as Einstein's theory of gravity), we must first figure out where the two theories intersect. Because of the smallness of Planck's constant h, the quantum theory deals with very tiny amounts, while Einstein's theory of gravity deals with very larger scale events, or at least with very huge masses that may drastically deform spacetime. The wave-particle duality is an essential feature of quantum theory; waves may behave as particles and particles may act as waves. Waves may also occur in the electromagnetic field, as has been demonstrated. Let us all compare electromagnetic fields to gravitational fields for the time being. The electric field seems smooth on a wide scale. Only at the microscopic level do we notice that the electric field is not uniform at all, but instead lumpy, since the electric field's energy is stored in minute bundles of electromagnetic energy called photons, rather than spread out in space. Likewise, according to quantum theory, the gravitational field must be quantized into small particles on a microscopic level, which we shall name gravitons (Deacon, 1981).

But what is a gravitational field if not spacetime warping? As a result, a quantum of gravity should also be a quantum of spacetime. In consequence, the graviton seems to be a spacetime quantum. As a result, spacetime itself could be presumably not smooth on a microscopic level, but rather has a graininess or bumpiness to it. No one knows for sure what happens to spacetime at this minuscule level, however, it has been hypothesized that spacetime could resemble a foam. When would quantum theory and Einstein's gravitational theory converge? Heisenberg's uncertainty principle provides the answer.

$\Delta E \Delta t \;\hbar$

Small quantities of energy ΔE of the electric field are converted into smaller quantities of energy, which are the photons, in the presence of an electric field. A similar process must be used to convert tiny quantities of gravitational field energy ΔE into small packets or quantum of gravity, known as gravitons, which may be stored in a computer. Taking into consideration the fact that a force's range is controlled by the mass of the exchanged particle, and that the gravitational force's range is known as infinite, it implies that the graviton's rest mass has to be zero. A quantum fluctuation should therefore manifest itself as a gravitational wave traveling at the speed of light c. Consequently, if we analyze the case of a spherically spreading gravitational field fluctuation, the little time required for it to traverse a distance is given by the expression:

$$\Delta t = \frac{r}{c} \quad (12)$$

We omit the larger than sign in the uncertainty principle to get an order of magnitude for the energy, and by putting Eqns. (12) into we obtain, for the energy of the oscillation,

$$\Delta E \Delta t = \Delta E \frac{r}{c} = \hbar$$

and,

$$\Delta E = \frac{\hbar c}{r} \qquad (13)$$

The value of r in Eqn. (13), that is where the quantum effects become significant, is uncertain at this moment; in fact, this is one of the things which we are attempting to figure out at this time. As a result, more information is required. Consider the amount of energy necessary for this little graviton or bundle of energy to move apart against the gravitational attraction of the surrounding space. Work done to separate the graviton is equivalent to the energy used to bring tiny pieces of the mass together from infinity. Let us first explore the issue in the electric field, and then apply the analogy to the gravitational field to see what we can learn from it. It is important to remember that the electric potential of a tiny spherical charge is:

$$V = \frac{kq}{r}$$

The electric potential V, on the other hand, was expressed as the potential energy (PE) per unit charge, such that,

$$V = \frac{PE}{q}$$

As a result, if a 2nd charge q is transferred from infinity to location r, the PE of the two-charge system is:

$$\text{PE} = qV = \frac{kq2}{r}$$

Similarly, the gravitational potential φ may have been calculated utilizing the same basic method as the electric potential. The gravitational potential will be as follows:

$$\varphi = \frac{GM}{r} \qquad (14)$$

where; *M* is the mass; *G* is the gravitational invariant; and *r* is the distance between the mass and the location at which the gravitational potential is to be calculated. Similarly, the electric potential, the gravitational potential of a spherical mass is expressed as the gravitational PE per unit mass. That is to say:

$$\varphi = \frac{PE}{M} \qquad (15)$$

As a result, bringing additional mass *M* from infinity to point *r* increases the *PE* of the system of two equal masses.

$$PE = M\varphi = \frac{GM2}{r} \qquad (16)$$

This value of the PE, required to arrange the two masses is the same energy required to separate them. When the same logic is applied to the assembly of masses that comprise the graviton, the PE supplied by Eqn. (16) equals the energy required to separate the graviton. The energy of the graviton, as determined by the uncertainty principle, may be equivalent to this energy. Therefore,

$PE = \Delta E$

By substituting ΔE from Eqn. (13) of the uncertainty principle and PE from Eqn. (16), we obtain:

$$\frac{GM2}{r} = \frac{\hbar C}{r} \qquad (17)$$

However, Einstein's mass-energy connection may be used to link the mass of the graviton *M* to the energy of the graviton as follows:

$$\Delta E = mc^2$$

or,

$$M = = \frac{E}{c2} \qquad (18)$$

Eqn. (18) is substituted into Eqn. (17) to obtain:

$$\frac{G(E)2}{r(c2)2} = \frac{\hbar c}{r}$$

Solving for *E*, we obtain:

$$E = \sqrt{\frac{\hbar c^5}{G}} \qquad (19)$$

The graviton's energy is represented by Eqn. (19). From the standpoint of particle physics, the graviton seems to be a particle with a mass of 10^{19} GeV/ c_2. When contrasted to the energies and masses of all the other fundamental particles, this is an enormous amount of energy and mass.

Quantum theory and gravity should be considered for any fundamental particles of this size or bigger. Remember that gravity was neglected in all other fundamental particle interactions. This energy is related to a mass of 2×10^{-5} g, which is an extremely small mass in terms of standard gravity.

REFERENCES

1. Aaltonen, T., Abazov, V. M., Abbott, B., Acharya, B. S., Adams, M., Adams, T., & Cordelli, M., (2018). Tevatron Run II combination of the effective leptonic electroweak mixing angle. *Physical Review D, 97*(11), 112007.
2. Abdel-Raouf, M. A., (1987). Are particles and antiparticles able to form (quasi) molecular structures?. In: *Few-Body Problems in Particle, Nuclear, Atomic, and Molecular Physics* (pp. 498–503). Springer, Vienna.
3. Abov, Y. G., Dzheparov, F. S., & Okun, L. B., (1984). On a verification of equality of particle and antiparticle masses in neutron-antineutron oscillations. *Pis' ma v Zhurnal Ehksperimental'noj i Teoreticheskoj Fiziki, 39*(10), 493–494.
4. Adamczyk, L., Adkins, J. K., Agakishiev, G., Aggarwal, M. M., Ahammed, Z., Alekseev, I., & LaPointe, S., (2013). Observation of an energy-dependent difference in elliptic flow between particles and antiparticles in relativistic heavy ion collisions. *Physical Review Letters, 110*(14), 142301.
5. Ahn, H. M., Lee, J. Y., Kim, J. K., Oh, Y. H., Jung, S. Y., & Hahn, S. C., (2011). Finite-element analysis of short-circuit electromagnetic force in power transformer. *IEEE Transactions on Industry Applications, 47*(3), 1267–1272.
6. Ahn, H. M., Oh, Y. H., Kim, J. K., Song, J. S., & Hahn, S. C., (2012). Experimental verification and finite element analysis of short-circuit electromagnetic force for dry-type transformer. *IEEE Transactions on Magnetics, 48*(2), 819–822.
7. Albers, M., Kiefer, C., & Reginatto, M., (2008). Measurement analysis and quantum gravity. *Physical Review D, 78*(6), 064051.
8. Aloisio, R., Galante, A., Grillo, A. F., Méndez, F., Carmona, J. M., & Cortés, J. L., (2004). Particle and antiparticle sectors in DSR1 and κ-Minkowski space-time. *Journal of High Energy Physics, 2004*(05), 028.
9. Araki, Y., Suenaga, D., Suzuki, K., & Yasui, S., (2021). Two relativistic kondo effects: Classification with particle and antiparticle impurities. *Physical Review Research, 3*(1), 013233.

10. Arkkio, A., Antila, M., Pokki, K., Simon, A., & Lantto, E., (2000). Electromagnetic force on a whirling cage rotor. *IEE Proceedings-Electric Power Applications, 147*(5), 353–360.

11. Azimov, Y. I., & Ioganssen, A. A., (1981). Difference between the partial decay widths of charmed particles and antiparticles. *Sov. J. Nucl. Phys. (Engl. Transl.);(United States), 33*(2).

12. Barnett, S. M., & Loudon, R., (2006). On the electromagnetic force on a dielectric medium. *Journal of Physics B: Atomic, Molecular and Optical Physics, 39*(15), S671.

13. Basov, N. G., Krokhin, O. N., & Popov, Y. M., (1961). Production of negative temperature states in pn junctions of degenerate semiconductors. *Sov. Phys. JETP, 13*(6), 1320, 1321.

14. Berger, J., Long, A. J., & Turner, J., (2019). Phase of confined electroweak force in the early universe. *Physical Review D, 100*(5), 055005.

15. Bernstein, J., & Feinberg, G., (1962). Electromagnetic mixing effects in elementary-particle physics. *Il Nuovo Cimento (1955–1965), 25*(6), 1343–1355.

16. Bock, F., (2019). *arXiv: Shining a Light on the QGP-Electroweak Probes Experimental Summary* (No. arXiv: 1901.10950, p. 011). SISSA.

17. Boulware, D. G., & Deser, S., (1975). Classical general relativity derived from quantum gravity. *Annals of Physics, 89*(1), 193–240.

18. Bourquin, M., Brown, R. M., Chatelus, Y., Chollet, J. C., Croissiaux, M., Degre, A., & Trischuk, J., (1979). Particle and antiparticle production by 200 GeV/c protons in the charged hyperon beam at the CERN SPS. *Nuclear Physics B, 153*, 13–38.

19. Brandenburg, G., Carnegie, R. K., Cashmore, R. J., Davier, M., Leith, D. W., Matthews, J., & Winkelmann, F., (1975). Measurement of particle and antiparticle elastic scattering on protons between 6 and 14 GeV. *Physics Letters B, 58*(3), 367–370.

20. Braun, D., (2001). Assets and liabilities are the momentum of particles and antiparticles displayed in Feynman-graphs. *Physica A: Statistical Mechanics and its Applications, 290*(3, 4), 491–500.

21. Bruschi, D. E., Dragan, A., Fuentes, I., & Louko, J., (2012). Particle and antiparticle bosonic entanglement in noninertial frames. *Physical Review D, 86*(2), 025026.

22. Budker, G. I., & Skrinskiĭ, A. N., (1978). Electron cooling and new possibilities in elementary particle physics. *Soviet Physics Uspekhi, 21*(4), 277.
23. Calcagni, G., (2020). Quantum gravity and gravitational-wave astronomy. *Handbook of Gravitational Wave Astronomy,* 1–27.
24. Carlip, S., (2008). Is quantum gravity necessary?. *Classical and Quantum Gravity, 25*(15), 154010.
25. Castell, L., (1966). Analysis of space-time structure in elementary particle physics. *Il Nuovo Cimento A (1965–1970), 46*(1), 1–38.
26. Chaichian, M., Fujikawa, K., & Tureanu, A., (2012). Lorentz invariant CPT violation: Particle and antiparticle mass splitting. *Physics Letters B, 712*(1, 2), 115–118.
27. Chaumet, P. C., & Nieto-Vesperinas, M., (2000). Coupled dipole method determination of the electromagnetic force on a particle over a flat dielectric substrate. *Physical Review B, 61*(20), 14119.
28. Chaumet, P. C., & Rahmani, A., (2009). Electromagnetic force and torque on magnetic and negative-index scatterers. *Optics express, 17*(4), 2224–2234.
29. Childs, G. N., Brand, S., & Abram, R. A., (1986). Intervalence band absorption in semiconductor laser materials. *Semiconductor Science and Technology, 1*(2), 116.
30. Chiu, H. Y., (1966). Symmetry between particle and antiparticle populations in the universe. *Physical Review Letters, 17*(13), 712.
31. Coldren, L. A., Fish, G. A., Akulova, Y., Barton, J. S., Johansson, L., & Coldren, C. W., (2004). Tunable semiconductor lasers: A tutorial. *Journal of Lightwave Technology, 22*(1), 193.
32. Coleman, J. J., (2012). The development of the semiconductor laser diode after the first demonstration in 1962. *Semiconductor Science and Technology, 27*(9), 090207.
33. Cremaschini, C., & Tessarotto, M., (2013). Symmetry properties of the exact EM radiation-reaction equation for classical extended particles and antiparticles. *International Journal of Modern Physics A, 28*(18), 1350086.
34. Crowley, M. T., Naderi, N. A., Su, H., Grillot, F., & Lester, L. F., (2012). GaAs-based quantum dot lasers. In: *Semiconductors and Semimetals* (Vol. 86, pp. 371–417). Elsevier.

35. Daneu, V., DeGloria, D. P., Sanchez, A., Tong, F., & Osgood, Jr. R. M., (1986). Electron-pumped high-efficiency semiconductor laser. *Applied Physics Letters, 49*(10), 546–548.

36. De Aquino, F., (2002). *Mathematical Foundations of the Relativistic Theory of Quantum Gravity*. arXiv preprint physics/0212033.

37. De Loach, B. C., Hakki, B. W., Hartman, R. L., & D'asaro, L. A., (1973). Degradation of CW GaAs double-heterojunction lasers at 300 K. *Proceedings of the IEEE, 61*(7), 1042–1044.

38. Deacon, D. A., (1981). Basic theory of the isochronous storage ring laser. *Physics Reports, 76*(5), 349–391.

39. Drell, S. D., (1977). Elementary particle physics. *Daedalus,* 15–31.

40. Engel, A., & Friedrichs, R., (2002). On the electromagnetic force on a polarizable body. *American Journal of Physics, 70*(4), 428–432.

41. Eseev, M. K., & Meshkov, I. N., (2016). Traps for storing charged particles and antiparticles in high-precision experiments. *Physics-Uspekhi, 59*(3), 304.

42. Fredsted, J., (2007). *Linking Electroweak and Gravitational Generators*. arXiv preprint arXiv:0712.2435.

43. Fredsted, J., (2010). *Electroweak Interaction Without Projection Operators Using Complexified Octonions.* arXiv preprint arXiv:1011.5633.

44. Gatti, E., & Manfredi, P. F., (1986). Processing the signals from solid-state detectors in elementary-particle physics. *La Rivista del Nuovo Cimento (1978–1999)*, *9*(1), 1–146.

45. Glashow, S. L., (1980). The future of elementary particle physics. In: *Quarks and Leptons* (pp. 687–713). Springer, Boston, MA.

46. Gleeson, A. M., & Morley, P. D., (1992). Generalized electroweak supercurrents in nuclear matter. *Nuclear Physics B, 369*(3), 655–665.

47. Goldman, T., Hughes, R. J., & Nieto, M. M., (1986). Experimental evidence for quantum gravity?. *Physics Letters B, 171*(2, 3), 217–222.

48. Gorelik, G. E., (1992). The first steps of quantum gravity. *Studies in the History of General Relativity, 3*, 367.

49. Gorelik, G. E., (2005). Matvei Bronstein and quantum gravity: 70th anniversary of the unsolved problem. *Physics-Uspekhi, 48*(10), 1039.

50. Hooft, G. T., (1993). *Dimensional Reduction in Quantum Gravity*. arXiv preprint gr-qc/9310026.

51. Isham, C., (1995). *Structural Issues in Quantum Gravity.* arXiv preprint gr-qc/9510063.
52. Kazakov, D. I., (2019). Prospects of elementary particle physics. *Physics-Uspekhi, 62*(4), 364.
53. Kieu, T. D., (1994). *Electroweak Interactions on the Lattice (No. UM-P--94/74).* Melbourne Univ.
54. Kim, S. J., & Lee, C. M., (2000). Investigation of the flow around a circular cylinder under the influence of an electromagnetic force. *Experiments in Fluids, 28*(3), 252–260.
55. Knudsen, H., & Reading, J. F., (1992). Ionization of atoms by particle and antiparticle impact. *Physics Reports, 212*(3, 4), 107–222.
56. Kopylov, G. I., & Podgoretskij, M. I., (1975). *Interference of Two-Particle States in Elementary Particle Physics and Astronomy* (No. JINR-R—2-8512). Joint Inst. for Nuclear Research.
57. Kopylov, G. I., & Podgoretsky, M. I., (1975). Interference of two-particle states in elementary-particle physics and astronomy. *Sov. Phys. JETP, 42*, 211–214.
58. Kriske, R., (2011). The electroweak force may be a result of a horizon of a curved universe. In: *APS Northwest Section Meeting Abstracts* (Vol. 13, pp. C1–015).
59. Kriske, R., (2011). The horizon of the universe could be the source of the electroweak force. In: *APS New England Section Fall Meeting Abstracts* (pp. F4–007).
60. Kumar, A., & DebRoy, T., (2003). Calculation of three-dimensional electromagnetic force field during arc welding. *Journal of Applied Physics, 94*(2), 1267–1277.
61. Lee, T. D., (1991). At present, we have QCD for the strong force, the SU (2)* U (1) standard model for the electroweak force and general relativity for the gravitational force. *Together; Physics Up to 200 TeV, 28*, 73.
62. Lin, F., Zuo, S., Deng, W., & Wu, S., (2016). Modeling and analysis of electromagnetic force, vibration, and noise in permanent-magnet synchronous motor considering current harmonics. *IEEE Transactions on Industrial Electronics, 63*(12), 7455–7466.
63. Luo, J. W., Bester, G., & Zunger, A., (2009). Full-zone spin splitting for electrons and holes in bulk GaAs and GaSb. *Physical Review Letters, 102*(5), 056405.

64. Mansuripur, M., (2008). Electromagnetic force and torque in ponderable media. *Optics Express, 16*(19), 14821–14835.

65. Mansuripur, M., Zakharian, A. R., & Wright, E. M., (2013). Electromagnetic-force distribution inside matter. *Physical Review A, 88*(2), 023826.

66. Martín-Martínez, E., & Fuentes, I., (2011). Redistribution of particle and antiparticle entanglement in noninertial frames. *Physical Review A, 83*(5), 052306.

67. Mason, S. F., (1991). Origins of the handedness of biological molecules. *Biological Asymmetry and Handedness, 162*, 3.

68. Mattingly, J., (2005). Is quantum gravity necessary?. In: *The Universe of General Relativity* (pp. 327–338). Birkhäuser Boston.

69. Maxwell, J. C., (1868). XXVI. On a method of making a direct comparison of electrostatic with electromagnetic force; with a note on the electromagnetic theory of light. *Philosophical Transactions of the Royal Society of London*, (158), 643–657.

70. Meiman, N. N., (1964). On asymptotic equality of the differential cross sections for particles and antiparticles. *Zh. Eksperim. i Teor. Fiz., 46*.

71. Nagashima, Y., (2011). *Elementary Particle Physics: Quantum Field Theory and Particles V1* (Vol. 1). John Wiley & Sons.

72. Okun, L. B., (1998). Current status of elementary particle physics. *Physics-Uspekhi, 41*(6), 553.

73. Omnès, R., (1972). The possible role of elementary particle physics in cosmology. *Physics Reports, 3*(1), 1–55.

74. Pangilinan, M., (2010). *Top Quark Produced Through the Electroweak Force: Discovery Using the matrix Element Analysis and Search for Heavy Gauge Bosons Using Boosted Decision Trees* (Doctoral dissertation, Brown University).

75. Passon, O., Zügge, T., & Grebe-Ellis, J., (2018). Pitfalls in the teaching of elementary particle physics. *Physics Education, 54*(1), 015014.

76. Peskin, M. E., (2008). Supersymmetry in elementary particle physics. In: *Colliders and Neutrinos: The Window into Physics Beyond the Standard Model (TASI 2006)* (pp. 609–704).

77. Plotnitsky, A., (2021). "Something happened:" on the real, the actual, and the virtual in elementary particle physics. *The European Physical Journal Special Topics, 230*(4), 881–901.

78. Porsev, S. G., Beloy, K., & Derevianko, A., (2009). Precision determination of electroweak coupling from atomic parity violation and implications for particle physics. *Physical Review Letters, 102*(18), 181601.

79. Portelli, C., (1987). The biomolecular mechanisms of motion and the role of the electroweak force. *Physiologie (Bucarest), 24*(3), 205–208.

80. Qiu, L., Li, Y., Yu, Y., Abu-Siada, A., Xiong, Q., Li, X., & Cao, Q., (2019). Electromagnetic force distribution and deformation homogeneity of electromagnetic tube expansion with a new concave coil structure. *IEEE Access, 7*, 117107–117114.

81. Qiu, L., Yi, N., Abu-Siada, A., Tian, J., Fan, Y., Deng, K., & Jiang, J., (2020). Electromagnetic force distribution and forming performance in electromagnetic forming with discretely driven rings. *IEEE Access, 8*, 16166–16173.

82. Recami, E., & Rodrigues, Jr. W. A., (1985). Tachyons: May they have a role in elementary particle physics?. *Progress in Particle and Nuclear Physics, 15*, 499–517.

83. Reifler, F., & Morris, R., (1985). A prediction of the Cabibbo angle in the vector model for electroweak interactions. *Journal of Mathematical Physics, 26*(8), 2059–2066.

84. Rovelli, C., (2003). Loop quantum gravity. *Physics World, 16*(11), 37.

85. Salam, A., (1979). The electroweak force, grand unification and superunification. *Physica Scripta, 20*(2), 216.

86. Skrinskiĭ, A. N., (1982). Accelerator and detector prospects of elementary particle physics. *Soviet Physics Uspekhi, 25*(9), 639.

87. Smarandache, F., (2009). A new form of matter—unmatter, composed of particles and anti-particles. *Neutrosophic Logic, Wave Mechanics, and Other Stories,* 8.

88. Smolin, L., (2010). *Newtonian Gravity in Loop Quantum Gravity.* arXiv preprint arXiv:1001.3668.

89. Szczekowski, M., (1989). Diquarks in elementary particle physics. *International Journal of Modern Physics A, 4*(16), 3985–4035.

90. Takatsu, N., KATO, M., Sato, K., & Tobe, T., (1988). High-speed forming of metal sheets by electromagnetic force. *JSME International Journal; Ser. 3, Vibration, Control Engineering, Engineering for Industry, 31*(1), 142–148.

91. THOMAS, A., (2006). Precision electroweak studies: An essential component of the worldwide nuclear physics program. In: *Honor of D Allan Bromley—Nuclear Scientist and Policy Innovator* (pp. 92–101).
92. Tomura, S., & Kunieda, M., (2009). Analysis of electromagnetic force in wire-EDM. *Precision Engineering, 33*(3), 255–262.
93. Toussaint, D., & Wilczek, F., (1983). Particle–antiparticle annihilation in diffusive motion. *The Journal of Chemical Physics, 78*(5), 2642–2647.
94. Wang, W., Yi, F., Ni, Y., Zhao, Z., Jin, X., & Tang, Y., (2000). Parity violation of electroweak force in phase transitions of single crystals of D-and L-alanine and valine. *Journal of Biological Physics, 26*(1), 51–65.
95. Wesson, P., (2013). *Gravity, Particles, and Astrophysics: A Review of Modern Theories of Gravity and G-Variability, and Their Relation to Elementary Particle Physics and Astrophysics*.(pp. 1-25)
96. Xiong, Q., Tang, H., Wang, M., Huang, H., Qiu, L., Yu, K., & Chen, Q., (2019). Design and implementation of tube bulging by an attractive electromagnetic force. *Journal of Materials Processing Technology, 273*, 116240.
97. Yoshikawa, T., & Morita, K., (2005). Refining of Si by the solidification of Si-Al melt with electromagnetic force. *ISIJ International, 45*(7), 967–971.
98. Zakharian, A. R., Mansuripur, M., & Moloney, J. V., (2005). Radiation pressure and the distribution of electromagnetic force in dielectric media. *Optics Express, 13*(7), 2321–2336.
99. Zel'dovich, Y. B., & Khlopov, M. Y., (1981). The neutrino mass in elementary-particle physics and in big bang cosmology. *Soviet Physics Uspekhi, 24*(9), 755.
100. Zichichi, A., (1979). New developments in elementary particle physics. *La Rivista del Nuovo Cimento (1978–1999), 2*(14), 1–65.

CHAPTER 5

NUCLEAR PHYSICS

CONTENTS

5.1. INTRODUCTION

Henri Becquerel (1852–1908) discovered in 1896 that uranium-containing ore generates hidden rays which can permeate paper as well as reveal a photographic plate. Upon Becquerel's exploration, Marie (1867–1934) as well as Pierre (1859–1906) Curie found polonium and radium, two additional radioactive components. Numerous experiments conducted by the Curies on such new pieces revealed that their radioactive material was unchanged through any physical or chemical process. By interacting with atomic electrons, substances produce chemical influence. The absence of chemical variations affecting radioactivity inferred that the orbital electrons have little to do with radioactive material (Bethe, 1937). Therefore, the radiation levels must originate from the nucleus. Rutherford studied this hidden radiation emanating from the nucleus of an atom by allowing it to keep moving in a magnetic field at right angles to the paper, as depicted in Figure 5.1. A few atoms were able to bend outwards, others have been bent straight down, and others were unbent through a magnetic field instead of being ready to bend in any way (Livingston and Bethe, 1937; Kondev et al., 2021).

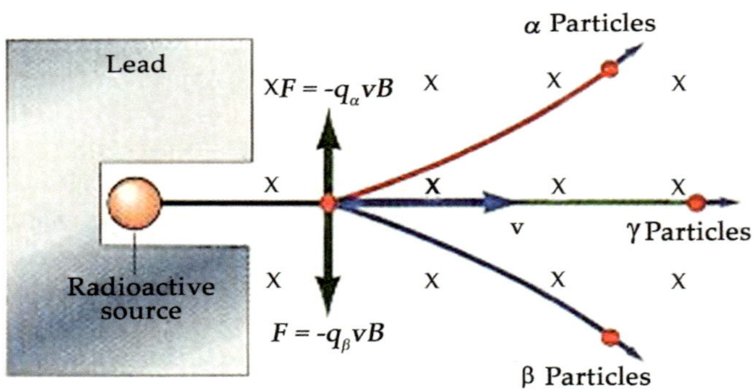

Figure 5.1. Radioactive particles.

Source: http://hyperphysics.phy-astr.gsu.edu/hbase/Nuclear/radact.html.

The atoms that have been curved up instead of down were referred to as alpha particles, those who were curved straight down as beta particles, and those who were not curved as gamma particles γ (Bethe and Bacher, 1936; Gal et al., 2016).

In overall physics, we learned that the resultant force substitute on a free electron affecting with velocity v in the magnetic field is provided as:

$F = qvB \sin\theta$

(Recollect that the right-hand rule is used to find the path of the magnetic force. Placing one's right hand in the line of the velocity vector v is a common way to demonstrate the direction of motion. Your thumb shows the direction of the force F operating on the molecule as you rotate your hand to the magnetic field B.) If the charge of the atom is positive, as seen in Figure 5.1, the attraction will act upwards, and the particle will be pushed up instead of down. Since the atom is seen to go up, it must have a positive electrical charge. It was discovered that its amplitude was double that of the electric charges. (It was eventually revealed that the particle is the nuclei of the diatomic molecule.)

Since the particle is pushed down by the magnetic field, it should be negatively charged. (It was discovered that particles are high-energy electrons. The absence of deflection of the particle in the magnetic field revealed that the particle has no electrical charge. Then, it has been discovered that particles are very energetic photons (Thompson, 1988; Koning and Rochman, 2008).

The energy of particles ranges between 0.1 MeV and 10 MeV, while the energy of outer electrons is measured in volts. In addition, it was discovered that electrons could scarcely through paper, but particles could enter several mm of aluminum and rays might reach many cm of leads. Thus, this provided more support for the notion that such cosmic rays must originate from the nucleus (Bohigas and Weidenmuller, 1988; Carlson et al., 2015).

5.2. NUCLEAR STRUCTURE

Following the satisfactory explanation of the characteristics of the atom by quantum mechanics, the next issue was, "What is the origin as well as the composition of the nucleus?" What is the arrangement of protons as well as neutrons in the nucleus? Why couldn't the nucleus disintegrate due to the particles' repulsive force? Are there electrons in the nuclei if the electrons which flow out of it are electrons? These issues will be discussed soon. Protons and neutrons form the nucleus. Nucleons are the aggregate name for protons and neutrons (Baldin, 1980; Bergman et al., 2007). The atomic number Z represents the number of protons in the nucleus, while the mass number A denotes the amount of neuronal (neutrons and protons) in nuclei. The neutron number N is the differential among the mass number as well as the atomic number, i.e., the number of particles in a nucleus,

$$N = A - Z \qquad (1)$$

A nucleus is metaphorically characterized by the type:

$$^{A}_{Z}X \quad (2)$$

Along with the mass number A superscripted and the atomic number Z subscripted, or at which X is the nuclei of the chemical substance represented by the atomic number Z. For instance, consider the notation " $^{12}_{6}C$ " symbolizes the nuclei of a carbon atom with an atomic number of 6, showing that it contains 6 protons, and a mass number of 12, showing that the nucleus contains 12 protons and neutrons. Eqn. (1) gives the number of neutrons, i.e.,

$$N = A - Z = 12 - 6 = 6$$

Isotopes may be discovered in every chemical component. The number of protons in an isotope of a chemical component is like the element's, however, the number of neutrons is unique. As a result, a chemical element's isotopes have a similar atomic number Z but a distinct mass number A as well as neutron number N (Ohlsen, 1972; Measday, 2001).

Because the amount of subatomic particles determines an element's chemical characteristics, an isotope has a similar number of electrons as the parent node and thus interacts thermally in a similar manner. The sole chemical variation is its atomic mass, which is determined by the presence or absence of nucleons (Brown, 1987; Blaum et al., 2013).

The carbon isotope is an instance of an isotope.

$$^{14}_{6}C$$

It contains the similar 06 protons as the parent node but possesses 14 nucleons, implying that now there are 14 – 6 = 8 neutrons There are three forms of hydrogen since the basic atom, hydrogen, contains two isotopes:

$^{1}_{1}H$ – Hydrogen has one proton and no neutrons.

$^{2}_{1}H$ – Deuterium is made up of one proton and one neutron.

$^{3}_{1}H$ – Tritium is made up of one proton and two neutrons.

There are two or more steady isotopes for almost every element. Isotopes are found in almost every chemical sample. The mean of the mass of the individual isotopes makes up an element's atomic mass. The number of isotopes of a given element is generally fairly low. Deuterium, for instance, is barely 0.015% abundant. As a result, the real atomic mass is rather near to A. Nevertheless, there are some outliers, one of which becomes the chlorine atom chlorinated. The atomic mass of chlorine is 35.5, adjusted to 03 significant numbers, as shown in the periodic chart. ${}^{35}_{17}Cl$ and ${}^{37}_{17}Cl$ are found in the chlorine test (Meißner, 2015; Durante and Paganetti, 2016).

${}^{35}_{17}Cl$ has a 75.5% richness, while ${}^{37}_{17}Cl$ has a 24.5% abundance. The mean of these two types of chlorine, balanced by the quantity of every contained in a material, is the atomic mass of chlorine. As a result, chlorine's atomic mass is:

Atomic mass = 35(0.755) + 37(0.245) = 35.5

A1(% Abundance) + A2(% Abundance) + A3(% Abundance) + atomic mass = A1(% Abundance) + A2(% Abundance)

wherein; A1, A2, and A3 are the mass numbers of each isotope.

Because the protons in a nucleus are polarized, Coulomb's law dictates that they repel each other, causing the nucleus to disintegrate. We infer there has to be some force inside the nucleus keeping these protons united since the nucleus does not blast them apart. The high nuclear force, or possible interactions, is the name given to this force of the nucleus. The strong force ties the nucleus combine because it works not just on protons but on neutrons. The reach of the powerful force is quite limited. That is, it functions inside a range of 10–14 m, which is on the scale of the nucleus' diameter.

There is no sign of this power outside the nucleus. The greatest force discovered is the strong nuclear force. We get a chart like the one in Figure 5.2 if we graph the neutrons in a nuclear N versus the number of protons in the same similar nucleus Z for multiple nuclei. The amount of neutrons in light nuclei is almost identical (Thomas, 1984; Engel et al., 1992).

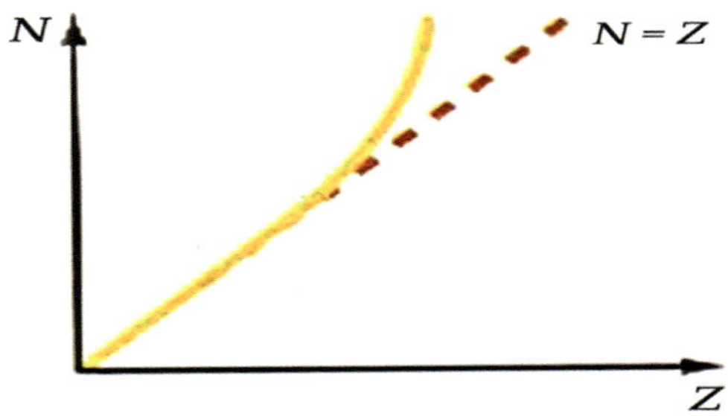

Figure 5.2. Graph of N v/s Z of atomic nuclei.

Source: http://physicsnet.co.uk/a-level-physics-as-a2/radioactivity/nuclear-instability/.

The line labeled N = Z represents the number of protons. There are many neutrons in the nuclei as compared to the protons as the atomic number Z grows. Keep in mind that the repulsive electrostatic attraction only exists among protons, whereas the strong nuclear attraction force exists among neutrons and protons. As a result, the extra neutrons enhance the force of attraction while decreasing the electrostatic repulsion and the core becomes more stable as a result of the electrostatic charge. The nucleus is steady once the nuclear attraction force is higher as compared to the electrostatic repulsive force. The nucleus splits apart or decomposes when the nuclear force will be lower as compared to the electrostatic force, emitting radioactive substances (Hodgson, 1971; Weidenmüller and Mitchell, 2009).

Chemical elements having Z values higher than 83 have radioactive atoms that disintegrate. In a similar fashion to Rutherford scattering, the interior composition of the nucleus is revealed. High-energy electrons (many 100 mega photon energies) assault the nucleus, penetrating it and reacting energetically with the proton inside it. As a consequence of this dispersion, protons, and neutrons appear to be spread very uniformly all across the nuclei, and the core is usually spherical or elliptical (Casten and Casten, 2000; Grawe et al., 2007).

All of it is commonly expected to only ever match the combination of its components. In the nuclei, however, that is not the case. The weight of the nuclei always is smaller as compared to the total mass of all neutrons and protons which make up the nuclear core, according to the findings of tests on the masses of various nuclei. The lost mass in the nuclei is known as the mass defect, m.

$$\Delta m = Z_{mp} + (A - Z)_{mn} - m_{nucleus} \quad (3)$$

Z_{mp} is the overall mass of all protons since Z is the overall number of protons and 'mp' is the mass of a proton. A–Z is the total number of neutrons, and because mn represents the mass of one neutron, $(A–Z)_{mn}$ is the mass of all neutrons, as indicated in Eqn. (1). The word nucleus refers to the mass of the whole nucleus as determined empirically. As a result, Eqn. (3) indicates the mass difference between the sum of its component masses plus the weight of the nucleus altogether (Hofstadter, 1956; Zink et al., 2004).

The remaining mass is turned into energy during the nucleus' creation. Einstein's mass-energy relationship yields this energy:

$$E = (\Delta m)c^2 \quad (4)$$

This is known as the nucleus' binding energy (BE). Eqns. (3) and (4) show that:

$$BE = (\Delta m)c^2 = Zm_pc^2 + (A - Z)\, m_nc^2 - m_{nucleus}c^2 \quad (5)$$

5.3. RADIOACTIVE DECAY LAW

Radioactivity refers to the sudden release of radiation from atomic nuclei. Radioactivity is the outcome of non-steady nuclei decaying or disintegrating. Radioactivity exists in nature in all chemical components with atomic numbers larger as compared to the 83 and several isotopes of chemical components with atomic numbers lower than 83. Several can be synthesized from almost all chemical components (Figure 5.3) (Frankfurt and Strikman, 1988; Kulagin and Petti, 2006).

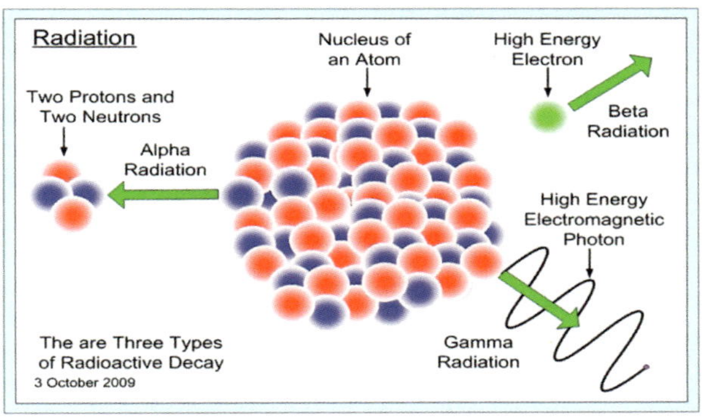

Figure 5.3. Radioactive decay.

Source: https://www.toppr.com/guides/physics/nuclei/radioactivity-law-of-radioactive-decay/.

The radioactive decay law measures the rate of radioactive emission. The amount of disintegrating nuclei dN throughout a specific time interval dt is dependent on the number of nuclei existing. Which is,

$$\frac{dN}{dt} \infty - \mathrm{N}$$

To get this comparison, we insert the proportionality constant, also known as the decay constant or disintegration constant, and we achieve an equality:

$$\frac{dN}{dt} \infty - \lambda \mathrm{N} \qquad (6)$$

The minus symbol is required in Eqn. (6) since the ultimate amount of nuclei N_f has always been lower than the starting amount of nuclei Ni; therefore, $dN = N_f - N_i$ always is negative, since there is always fewer radioactive nucleus over time (Haxel et al., 1949; Bohr and Mottelson, 1998). The constant of decay is dependent on the specific isotope of the chemical substance. A high value implies a high rate of loss, while a low ratio suggests a low rate of decay. The variable -dN/dt in Eqn. (6) is indeed the speed of decay of nuclei with duration, commonly known as the action and denoted by the letter A,

$$\mathrm{A} = -\frac{dN}{dt} = \lambda \mathrm{N} \qquad (7)$$

As to what proceeds, we have to be cautious not to mistake the sign A for action with the sign A for mass number. It must always be evident in the situation in which it is used (Butt and Wilson, 1972; Cooper, 2009).

The Standard unit of activity is the becquerel, which corresponds to single decay per second. That is to say,

1 Bq = 1 decay/s

An earlier measure of activity, the curie, written Ci, is equal to 3.7×10^{10} Bq per unit. The millicurie (10^{-3} curie = mCi) and the microcurie (10^{-6} curie = Ci) are discrete components of activity (Arthur et al., 1981; Gopych and Zalyubovskii, 1988).

The overall amount of nuclei available at each point in time is calculated by averaging Eqn. (6) from t = 0 to a time t as:

$$\frac{dN}{d} = -\lambda N$$

$$\int_{N^\circ}^{N} \frac{dN}{N} = -\int_{0}^{t} dt$$

Once t = 0, the lesser constrain on the right side, the original estimate of nuclei is N_0, the lesser constrain on the left-hand side of the equation, and yet when t = t, the outer bound on the right side, the amount of nuclei existing is N, the outer bound on the left side (Pommé and Pelczar, 2020; Cruz-López and Espinosa-Paredes, 2022). When we integrate, we get:

$$\ln N \int_{N^\circ}^{N} = -t \int_{0}^{t}$$

$$\ln N - \ln N_0 = -\lambda t$$

$$\ln \frac{N}{No} = -\lambda t$$

Given that $e^{\ln x} = x$, we must now add e to either side of the equation to obtain x,

$$\frac{N}{No} = e^{-t}$$

After calculating for N, we get at:

$$N = N^\circ e^{-t} \quad (8)$$

Eqn. (8) is the equation of radioactive decay, and it yields the overall number of nuclei existing at any point in time t. N_0 is the number of nuclei available at time t = 0, which marks the beginning of measurements of nuclei. Figure 5.4 depicts a visualization of the radioactive decay rule, Eqn. (8). The curve indicates the number of remaining radioactive nuclei at time ‘t’ (Bakaç et al., 2011; Sitek and Celler, 2015). Finding the time required for 50% of the initial nucleus to decay yields an incredibly exciting number such that barely 50% of the initial nuclei remain.

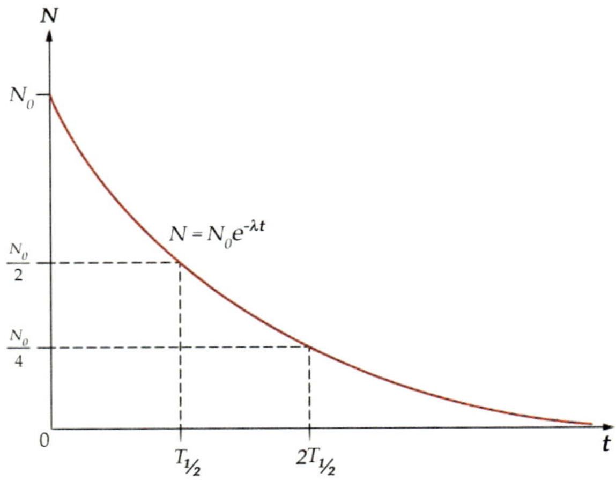

Figure 5.4. The radioactive decay law.

Source: https://opentextbc.ca/universityphysicsv3openstax/chapter/radioactive-decay/.

$N_0/2$ represents 50% of nuclei and is depicted in the image. The quantity of $N_0/2$ is shown by a horizontal line made till it crosses the curve N = N°e(–t). From such a position, a vertical line is lowered to the t-axis. The value of the t-axis is the time required for 50% of the initial nuclei to decay (De Marcillac et al., 2003). This duration measured along the t-axis is thus known as the ½ of the radioactive atoms and is indicated by T1/2. Therefore, the half-life of radioactive particles is the amount of time required for half of the initial radionuclides to decay. When t = T1/2, N = $N_0/2$, an essential connection among the ½ as well as the decay constant may be determined. Putting these numbers into the decay law in Eqn. (8) yields,

$$\frac{No}{2} = N^\circ e^{-T\,1/2}$$

$$\frac{1}{2} = e^{-T\,1/2}$$

The result of calculating the natural log of both sides of this equation is:

$$\ln\frac{1}{2} = \ln e^{-T\,½} \quad (9)$$

In contrast, the natural log ln is the reverse of the exponential curve e, so when given progressively, as on the right side of Eqn. (9), they canceled each other out and left the function alone. As a result, Eqn. (9) gets,

$$\ln\frac{1}{2} = T1/2$$

Calculating 0.693 using the normal logarithm ln of ½ on an electronic computer.

Thus,

$$-0.693 = -T1/2$$

Calculating the constant of decay,

$$\lambda = \frac{0.693}{T1/2} \quad (10)$$

Therefore, if we determine the half-life T1/2 of a radioactive nucleon, we may use Eqn. (10) to get its decay constant. Inversely, if we understand, we can determine the half-life using Eqn. (10) (Corral et al., 2011; Aston, 2012). Notice that the activity (i.e., the amount of disintegrations per second) is not constant, since it is dependent on N, which decreases with time according to Eqn. (8). If we insert Eqns. (8) into (7) for the action, we obtain:

$$A = \lambda N = \lambda No\, e^{T}$$

Letting,

$$\lambda No = A_o$$

The decay rate of the nuclei at time t = 0 is calculated for the action,

$$\mathrm{A} = \mathrm{A}_o e^{-T} \quad (11)$$

Recalling that the activity is the number of disintegrations per second, we see that, the rate of decay is not steady but continuously decreasing. Figure 5.5 depicts a graph of the activity as a function of the time (a) (Godovikov, 2004; Pommé and Pelczar, 2022). Take note of the resemblance between this diagram and Figure 5.4. For the period t, which is equal to the half-life, the activity is,

$$\mathrm{A} = \mathrm{A}_o e^{-T1/2}$$

Substituting from Eqn. (11), we get:

$$A = A_o e^{-[0.693\ T1/2]T1/2} = A_o e^{-0.693}$$

Using a calculator, we see that e(–0.693) = 0.500 = 12. Hence,

$$A = \frac{Ao}{2} \text{ for t} = T1/2 \qquad (12)$$

In other words, the decay rate is reduced by half during one half-life.

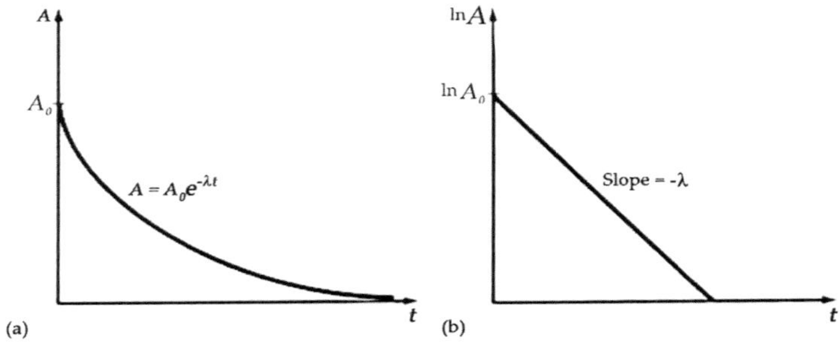

Figure 5.5. Radioactive activity.

Source: https://www.nuclear-power.com/nuclear-power/reactor-physics/atomic-nuclear-physics/radioactive-decay/activity-specific-activity/.

In experimental results, the decay constant may be determined using the formula. First, let us just go to Eqn. (11) and calculate the natural logarithm of either side, i.e.,

$$\ln A = \ln(A_o\ e^{-T})$$

The log of a product is equivalent to the total of the logarithms of every component, according to the rules for managing logarithms. Therefore,

$$\ln A = \ln A_o + \ln\ e^{-T}$$

However, as previously stated, the normal log, as well as the exponential, are opposites of one another, and so,

$$\ln\ e^{-T} = -\lambda t$$

Thus,

$$\ln A = \ln A_0 - \lambda t$$

Rearranging this results in,

$\ln A = -\lambda t + \ln A_o$ (13)

If we now walk inside the lab and check the number of disintegrations at unit time A at various periods t, we may plot ln A against t to produce the horizontal line seen in Figure 5.4(b). Since the gradient is –, maybe empirically verified. Once is known, we may calculate the half-life using Eqn. (13). Additionally, it is occasionally practical to utilize the average life or mean life T_{avg} of a sample (Huestis, 2002; Qi et al., 2009). The average or mean life is the average lifespan of all particles in a particular test piece. It ends out to be the reciprocal of the decay constant, or the decay constant itself,

$$T_{\mathrm{avg}} = \frac{1}{\lambda} \quad (14)$$

5.4. FORMS OF RADIOACTIVITY

Only the quantity of decaying nuclei has been described thus far, with no mention of the disintegration specifics. The decay of nuclei is possible:

- Alpha decay, α;
- Beta decay, β^-;
- Beta decay, β^+, positron emission;
- Electron capture;
- Gamma decay, γ.

Let us now go through each of them in further detail.

5.4.1. Alpha Decay

The electrostatic repulsive force begins to outweigh the nuclear attraction whenever a nucleus contains just so many protons in comparison to neutrons. Whenever it happens, the nucleus becomes non-steady and produces a radioactive decay particle (Coen, 2002; Jesse, 2003). Two protons and two neutrons are consequently lost from the nucleus. As a result, the quantity of protons in the atomic nucleus reduces by 2, whereas the quantity of protons and neutrons in the nucleus falls by 4. The nucleus is known as the "father" nucleus beforehand the decay and as the "daughter" nucleus ever since the decay (van Brakel, 1985; Krause et al., 2012). As a result, alpha decay is represented with the symbol,

$$^{A}_{z}X \rightarrow {}^{A-4}_{z-2}X + {}^{4}_{2}He \quad (15)$$

where; in ${}_{z}^{A}X$ is the parental nucleus, and ${}_{z-2}^{A-4}X$ is the baby nucleus, and ${}_{2}^{4}He$ is the α molecule, and is the helium nucleus. It is worth noting that the atomic number Z has shrunk by two components. This signifies that one compound with the atomic number Z has been transfigured into a chemical component with the atomic number Z-2 through alpha decay. The early alchemists aimed to transform chemical components, particularly base metals into gold (Weinert, 2009; Pfützner, 2013). Since they were dealing with chemical processes, that, as previously stated, is dependent on the electrical structure of atoms rather than their nucleus, this outcome never was achieved (Figure 5.6) (Langevin, 1985; Lee et al., 2021).

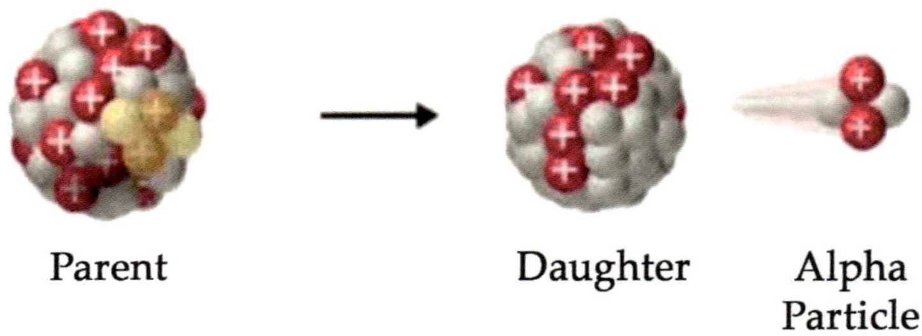

$$ {}_{Z}^{A}X \rightarrow {}_{Z-2}^{A-4}Y + {}_{2}^{4}He $$

Figure 5.6. Alpha decay.

Source: https://www.nuclear-power.com/nuclear-power/reactor-physics/atomic-nuclear-physics/radioactive-decay/alpha-decay-alpha-radioactivity/.

Uranium-238, which decays through photon discharge with a ½ of 4.51 × 10^9 years, is an instance of known natural alpha decay. Eqn. (15) is used to locate its daughter nucleus. Therefore,

$$ {}_{92}^{238}U \rightarrow {}_{90}^{234}X + {}_{2}^{4}He \qquad (16) $$

The atomic number Z has been reduced from 92 to 90. The chemical substance with the Z = 90 value is thorium, according to the periodic table (Konopinski and Uhlenbeck, 1941; Kielland-Brandt, 1974). As a result, uranium has been transfigured to thorium by particle discharge. Eqn. (16) has been changed to,

$$^{238}_{92}U \rightarrow ^{234}_{90}Th + ^{4}_{2}He \tag{17}$$

Also, notice as the mass number A for thorium in the atomic numbers is 232, but the mass number in Eqn. (17) is 234. This indicates the formation of a thorium isotope (Bretscher, 1940; Fadel, 2014). (In reality, $^{234}_{90}Th$ a volatile isotope that decays as well, but this time through beta radiation.) We will go into more detail about all this).

5.4.2. Beta Decay, β⁻

An electron is seen leaving the nucleus during beta decay. Nevertheless, according to the Heisenberg uncertainty principle, an electron cannot be confined in a nucleus. As a result, the electron must be produced inside the nucleus at the time of release. It has been discovered that a neutron inside the nucleus decays into a proton, an electron, and then the other molecule known as an antineutrino (Figure 5.7) (Davis et al., 1951; Smith, 1952).

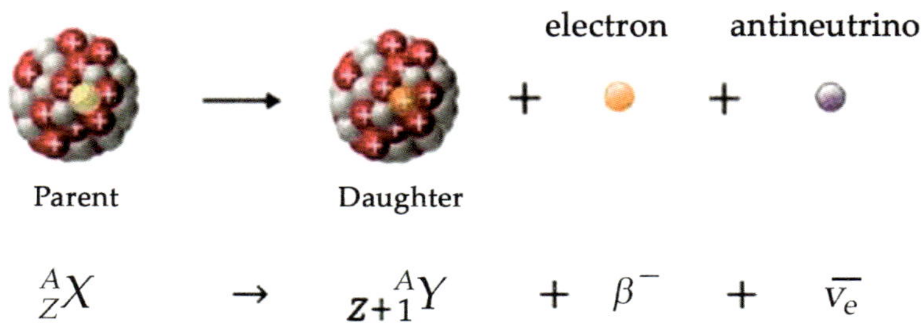

Figure 5.7. Beta decay.

Source: https://www.nuclear-power.com/nuclear-power/reactor-physics/atomic-nuclear-physics/radioactive-decay/beta-decay-beta-radioactivity/.

The Greek letter nu, v, along a bar on the v, which is v, is used to represent the antineutrino. The antineutrino is the neutrino's antiparticle. The neutron decay process is as stated,

$$^{1}_{0}n \rightarrow ^{1}_{1}p + ^{0}_{-1}e + \text{v} \tag{18}$$

The electron β^- or particle is denoted by the notation $^{0}_{-1}e$. It contains a mass number A of 0 since it lacks nucleons and an atomic number of −1 as it is not a positive molecule. Since it possesses mass number and atomic number one, the proton is represented as ${}_{1}^{1}p$. As a result, in beta decay, the nucleus drops a neutron and obtains a proton, whereas the particles, electrons, and antineutrino are expelled from the nucleus (Henderson et al., 1934; Doğru and Külahci, 2004). Therefore, the atomic number Z rises by one in the decay since the nucleus got a proton, however, the mass number A remains the same since, while losing one neutron, we acquired one proton.

A beta decay, β^-, may be expressed symbolically as,

$$ {}_{Z}^{A}X \rightarrow {}_{Z+1}^{A}X + {}_{-1}^{0}e + \text{v} \tag{19} $$

It should be noted that with beta decay, Z grows to Z + 1. As a result, a chemical component with the atomic number Z is transfigured into some other chemical substance with the atomic number Z + 1.

As an illustration, consider the isotope ${}_{90}^{234}Th$ is non-steady and decomposes through beta emission has a half-life of 24 days. Its decay may be expressed mathematically,

$$ {}_{90}^{234}Th \rightarrow {}_{91}^{234}X + {}_{-1}^{0}e + \text{v} \tag{20} $$

Staring up the periodic table, we notice that a chemical substance matching atomic number 91 is protactinium (Pa). As a result, the component thorium has been transfigured to the component protactinium. Also, according to the periodic table, the mass number A for protactinium ought to be 231. This is an isotope of protactinium because we contain a mass number of 234 (Tamplin, 1972; Mobaligh et al., 2021). (Anyone who is also volatile and decays.) The beta decay of thorium is currently expressed as:

$$ {}_{90}^{234}Th \rightarrow {}_{91}^{234}Pa + {}_{-1}^{0}e + \text{v} $$

5.4.3. Beta Decay ß$^+$ Positron Emission

A positron is released from the nucleus in this sort of disintegration. The positron is the electron's antiparticle. Except that it has a positive charge, it has all of the properties of an electron. Since there is no position in the nucleus, one must be formed beforehand it can be emitted (Soddy, 1906; Schmidt et al., 1955). The decomposition of a proton into a neutron, a positron, and a neutrino v causes positron emission, which is written symbolically as Eqn. (20).

Along with the neutrino ${}^{0}_{+1}$e the positron is released into the nuclei. In the nucleus, the neutron remains. As a consequence of the lack of the proton, the atomic number Z reduces by one in beta decay. Since a proton is managed to lose, a neutron is formed to maintain a similar number of protons and neutrons, and the mass number A remains constant (Studier and Hyde, 1948; Bozkurt et al., 2007). As a result, a ß+ decay can be documented as,

$${}^{A}_{Z}X \rightarrow {}^{A}_{Z-1}X + \quad +\text{v} \qquad (21)$$

For instance, the unsteady isotope of aluminum ${}^{26}_{13}Al$ decays by ß+ discharges with a half-life of 7.40 × 10^5 years. The equation is used to write down the reaction,

$${}^{26}_{13}Al \rightarrow {}^{26}_{12}X + {}^{0}_{+1}e \ +\text{v} \qquad (22)$$

The atomic number 12 relates to the chemical substance magnesium Mg, according to the elements in the periodic table,

$${}^{26}_{13}Al \rightarrow {}^{26}_{12}Mg + {}^{0}_{+1}e \ +\text{v}$$

We can tell that this transformation formed a magnesium isotope since the mass number A of magnesium is 24. It is vital to remember that proton breakdown (Eqn. (22)) can only happen inside the nucleus. Since this mass of the proton is smaller as compared to the neutron mass, a single proton could not decay into a neutron (English et al., 1947; Kirby, 1954).

5.4.4. Electron Capture

In the event, that an orbital electron approaches the nucleus as well closely and is sucked up by the nucleus. Because the electron could not exist inside the nucleus as an electron, it manages to combine with a proton to form a neutron as well as a neutrino. It is represented as,

$${}^{0}_{-1}e \ {}^{1}_{+1}p \rightarrow {}^{1}_{0}n \ +\text{v} \qquad (23)$$

This procedure reduces the number of protons in the nucleus by one, resulting in a change from Z to Z–1 while maintaining the number of nucleons A. Thus, this decay can be expressed as,

$${}^{0}_{-1}e + {}^{A}_{Z}X \rightarrow {}^{A}_{Z-1}X \ +\text{v} \qquad (24)$$

Whenever the electron closest to the nucleus is taken prisoner by the nucleus, the electron orbit becomes vacant. An electron from a space with a higher energy state enters this vacant position (Wahl, 1941; Hagemann et al., 1947). In the X-ray region of the electromagnetic spectrum, a photon is released when the difference in energy between the electron in the upper orbit and the electron in the ground orbit is greater than one (Figure 5.8) (Elliott and King, 1941).

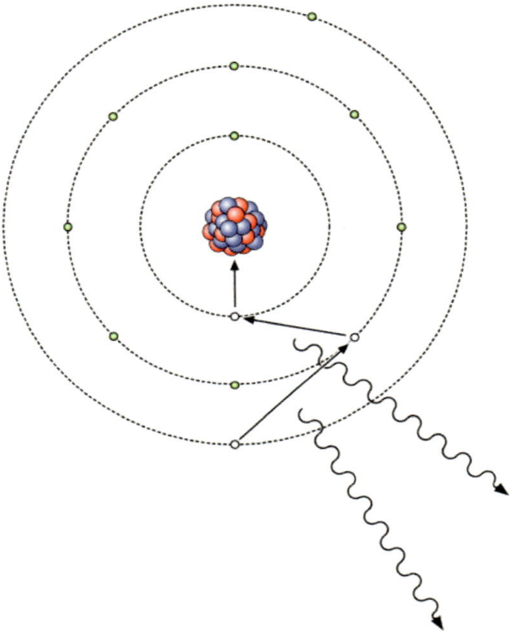

Figure 5.8. Graphical representation of electron capture.

Source: https://socratic.org/questions/how-does-electron-capture-work.

As an instance of gaining electrons, take into account the mercury isotope $^{197}_{80}Hg$, whose half-life is 65 hours due to electron capture. This decay can be symbolized using Eqn. (24) (Hagemann et al., 1950; Radenković et al., 2015).

$$^{0}_{-1}e + ^{197}_{80}Hg \rightarrow ^{197}_{79}X + \text{v}$$

According to the periodic chart, the atomic number Z = 79 corresponds to the alkali metal gold, Au.:

$$^{0}_{-1}e + ^{197}_{80}Hg \rightarrow ^{197}_{79}Au + \text{v}$$

Finally, the old alchemists' hopes have been realized. A mercury isotope has been transformed into the material gold. Also notice that gold's mass number A is 197. Therefore, the transformation has produced gold, a strong metal (Andrews, 2009; Patra et al., 2013).

5.4.5. Gamma Decay

A nucleus experiencing decay is occasionally left in an excited condition. As an element in an active state of such an atom releases a photon as well as falls back to ground level, a proton or neutron might be in an active state within the nucleus. Whenever the nucleon lowers back to the normal state, thus it releases a photon. Since the energy shut off is so great, the frequency of the photon would be in the gamma-ray section of an electromagnetic spectrum (Voltaggio et al., 2001; Papadopoulos et al., 2013). Therefore, the excited nucleus recovers to its lowest energy, and a gamma-ray is generated. Thus, gamma decay is demonstrated the idea as,

$${}_{Z}^{A}X^{*} \rightarrow {}_{Z}^{A}X + \gamma \quad (25)$$

where; the * just on the nucleus represents an exciton. In this form of decay, neither atomic number Z nor the mass number A alters. Therefore, gamma decay does not transform some of the chemical components (Figure 5.9) (Grosse et al., 1941; Ribeiro, 1998).

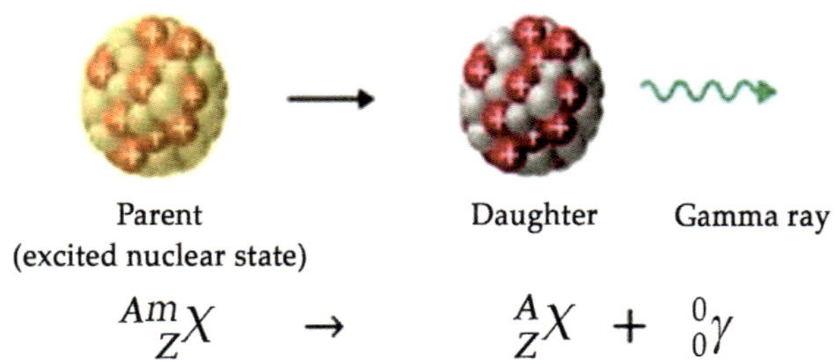

Figure 5.9. Graphical representation of gamma decay.

Source: https://www.nuclear-power.com/nuclear-power/reactor-physics/atomic-nuclear-physics/radioactive-decay/gamma-decay-gamma-radioactivity/.

5.5. RADIOACTIVE SERIES

Components with atomic numbers Z larger than 83 are unsteady and decompose normally, as previously stated. The majority of these unsteady elements have short lives and decay rapidly. As a result, they are uncommon naturally. The components Thorium-232, Uranium-238, and uranium isotope 235 are the only exclusions (Sakanoue, 1967; Adloff, 1981).

The element:

$^{232}_{90}Th$ has a half-life of 1.39×10^{10} yr

$^{238}_{92}U$ has a half-life of 4.50×10^{9} yr

and,

$^{235}_{92}U$ has a half-life of 7.10×10^{8} yr

Furthermore, these components devolve into a succession of daughters, granddaughters, great-granddaughters, and so forth. Figure 5.10 illustrates the series decay (_90232)Th, which would be a graph of the neutron number N vs the atomic number Z. (_90232)Th is displayed using the coordinates N = 142 and Z = 90 since it contains a Z value of 90 and an N value of 232 – 90 = 142. With a ½ of 1.39 × 10^{10} years, the first (_90232)Th decays through alpha emission (Rieppo, 1978; Fajans, 2013). An alpha decay lowers the atomic number Z to Z–2 and reduces the mass number A by 4 to A–4, as shown in Section 5.4, Eqn. (15). As a result, (_90232)Th decays as,

$$^{232}_{90}Th \rightarrow {}^{228}_{88}X + {}^{4}_{2}He$$

However, atomic number 88 is associated with the element radium (Ra). Hence,

$$^{232}_{90}Th \rightarrow {}^{228}_{88}Ra + {}^{4}_{2}He$$

228 – 88 = 140 is the neutron number N for (_88228)Ra. Therefore, the figure with parameters N = 140 and Z = 88 yields (_88228)Ra. The initial neutron number is calculated as follows:

$$N_0 = A - Z$$

However, in alpha radiation, A becomes A–4 and Z becomes Z–2, as seen in Eqn. (15). As a result, the new neutron number is $N_1 = (A - 4) - (Z - 2)$,

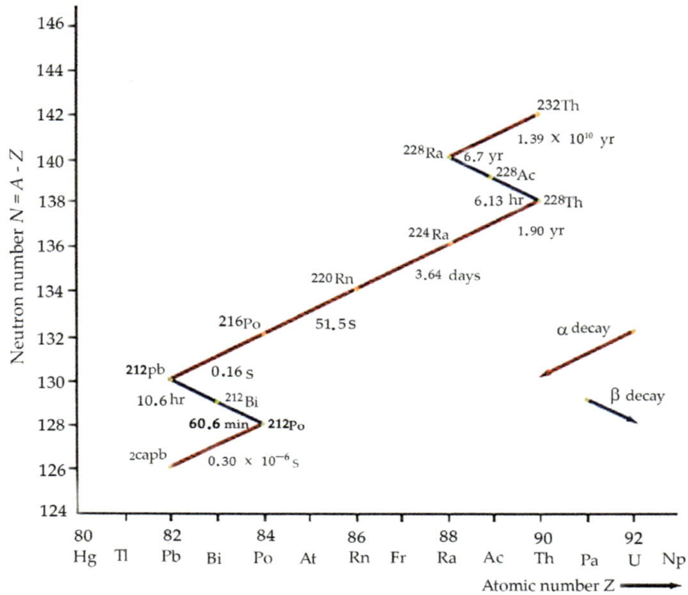

Figure 5.10. Thorium decay series.

Source: https://metadata.berkeley.edu/nuclear-forensics/Decay%20Chains.html.

$= A - Z - 2$

$N_1 = N_0 - 2$ (alpha decay) (26)

As a result, the neutron number reduces by 2 for all alpha releases. As a result, for each alpha emission in the figure, both N and Z have dropped by 2. Radium-228 is also unsteady and decomposes by beta discharges with a half-life of 6.7 yr. The atomic number Z grows to Z + 1 as given in Eqn. (26), but the mass number A stays unchanged (Vogt, 1962; Sargsian, 2001). The beta neutron number emission transforms into,

$N_1 = A - (Z + 1) = A - Z - 1$

$N_1 = N_0 - 1$ (beta-decay) (27)

As a result, (_88228)Ra produces (_88229)Ac, with N = 139 and Z = 89 coordinates. As a result, alpha emission shows as a line curving down to the left in the periodic graph, with either N and Z dropping by two units. In contrast, beta emission shows a downward sloping line to the right, with N dropping by 1 and Z rising by 1. Figure 5.10 depicts the whole family's degeneration thorium-232 decays by discharges to Radium-228, where it then decomposes by – emission to Actinium-228, that decays by – to Thorium-228, which then decomposes by discharges to Radium-224, which further decomposes by discharges to Radon-220, which again decomposes by discharges to Polonium-216, which therefore decomposes by discharges to Lead-212, where it then decomposes by – to Bismuth-212, which further decomposes α emission to the stable Lead-208. The figure depicts the half-life of each degradation (Weisskopf, 1937; Nifenecker and Pinston, 1990).

A series refers to the radioactive cycle. Figure 5.11 depicts the decay sequence for Uranium-238. It begins with (_38292)U and finishes with Lead-206, a stable isotope. The decay sequence for Uranium-235 is shown in Figure 5.12. The sequence comes to a close with the chemically stable component Lead-207. The neptunium family, which finishes in the chemically stable element Bismuth-209, is seen in Picture 5.13 (Serber, 1947; Hodgson, 1971). That it is above uranium in the periodic table, neptunium is considered a transuranic component. Uranium is the strongest chemical component found in nature, having an atomic number of Z = 92. Man has created elements with a Z higher than 92. These new additions may also be made into a variety of isotopes (Silberberg and Tsao, 1973; Sato and Yazaki, 1981).

The interaction of hitting (_38292)U with neutrons produces neptunium as an instance of the formation of a transuranic metal,

$$ {}^{292}_{38}U + {}^{1}_{0}n \rightarrow {}^{239}_{93}Np + {}^{0}_{+1}e + \text{v} \quad (28) $$

That is, (_38292)U receives the neutron but then emits an electron by beta decay. The atomic number is raised via one, between Z = 92 to Z = 93, resulting in the birth of a new chemical substance known as neptunium. As per the process, Neptunium-239 is volatile and decomposes through beta release, producing other chemical components called plutonium,

$$ {}^{239}_{93}Np \rightarrow {}^{239}_{94}Pu + {}^{0}_{-1}e + \text{v} \quad (29) $$

Americium was the second transuranic material to be produced, and it was generated by:

$$ {}^{239}_{94}Pu + {}^{1}_{0}n \rightarrow {}^{240}_{94}Pu + \gamma $$

$$ {}^{240}_{94}Pu + {}^{1}_{0}n \rightarrow {}^{241}_{94}Pu + \gamma $$

$$ {}^{241}_{94}Pu + {}^{1}_{0}n \rightarrow {}^{242}_{94}Pu + \gamma $$

Such that, neutrons are used to target plutonium till the isotope (_94241) Pu is formed, which subsequently beta decays to produce the isotope of the unique chemical component americium (Weisskopf and Ewing, 1940; Feshbach, 1958). More components were generated by colliding elements with different additional atoms and elements. We get some instances of the construction of new components as well,

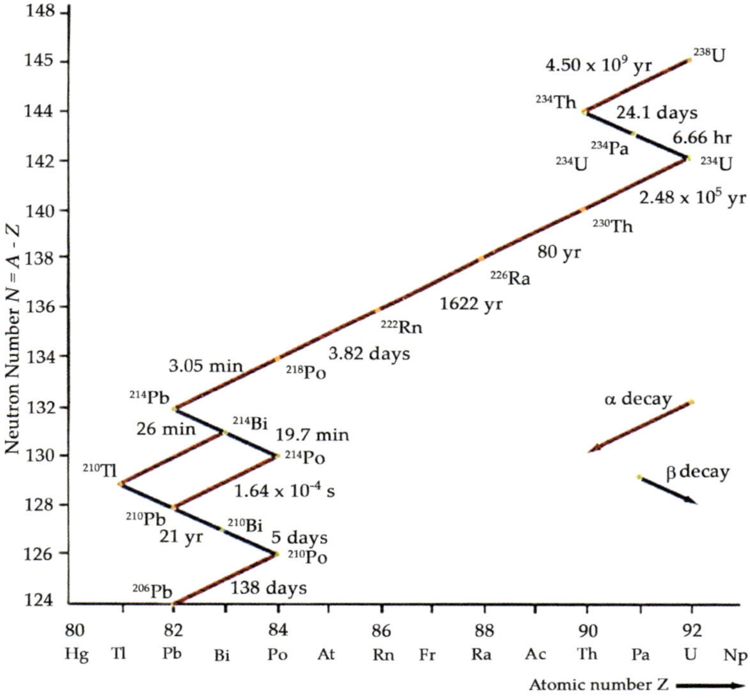

Figure 5.11. Uranium (U) decay series.

Source: https://www.researchgate.net/figure/Decay-series-of-A-uranium-238-238-U-B-uranium-235-235-U-and-C-thorium-232_fig4_313744940.

(Curium) $^{239}_{94}Pu + ^{4}_{2}He \rightarrow ^{242}_{96}Cm + ^{1}_{0}n$

(Berkelium) $^{241}_{94}Am + ^{4}_{2}He \rightarrow ^{243}_{97}Bk + ^{0}_{-1}e + v + 2^{1}_{0}n$

(Californium) $^{242}_{96}Cm + ^{4}_{2}He \rightarrow ^{245}_{98}Cf + ^{1}_{0}n$

(Nobelium) ${}^{246}_{96}Cm + {}^{12}_{6}C \rightarrow {}^{254}_{102}No + 4\,{}^{1}_{0}n$

(Lawrencium) ${}^{252}_{98}Cf + \mathrm{B} \rightarrow {}^{257}_{103}Lr + 5\,{}^{1}_{0}n$

Further research discovered that the neptunium decay sequence begins with plutonium (_94239)Pu, which decays via beta emission to americium (_94241)Am, which again decomposes by alpha transmission to neptunium (_93237)Np (Siegert, 1939; Strauch, 1951).

Since this primary component of these groups has such a lengthy lifespan, the majority of the series' members are discovered naturally. As several isotopes decompose as are being created, an equilibrium situation is achieved within the sequence. Man-made isotopes have been created. They most likely existed as well (Figure 5.13) (Miller and Hudis, 1959; Trautmann et al., 2010).

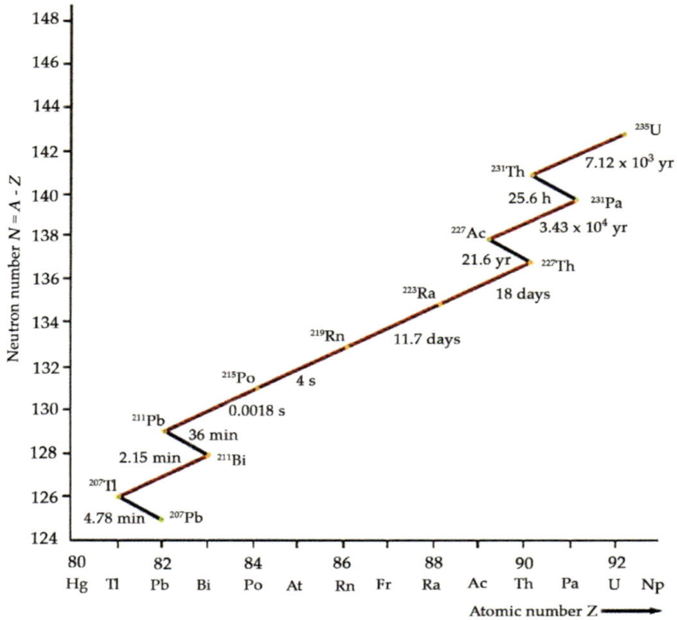

Figure 5.12. The neptunium series.

Source: https://metadata.berkeley.edu/nuclear-forensics/Decay%20Chains.html.

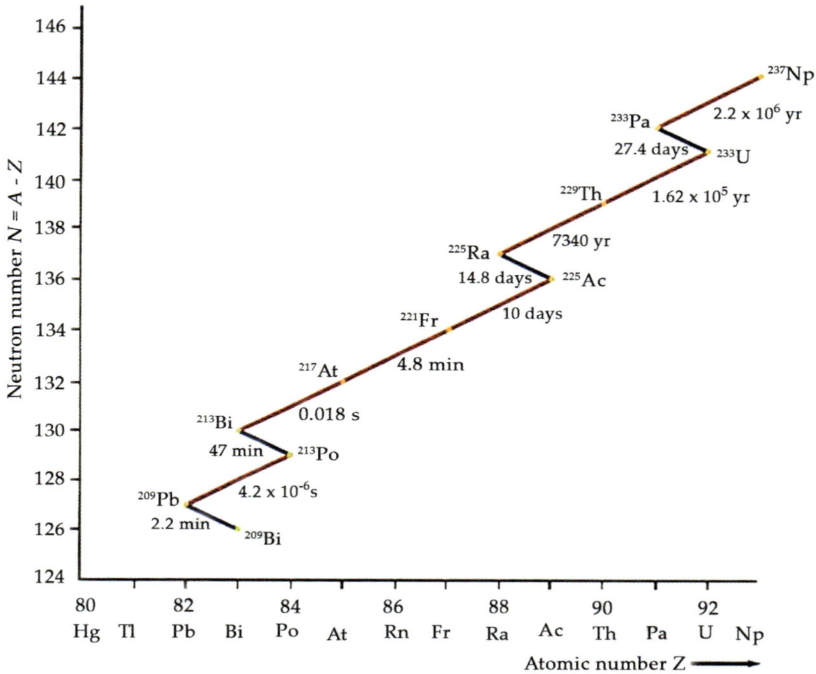

Figure 5.13. The neptunium series.

Source: https://commons.wikimedia.org/wiki/File:Decay_Chain(4n%2B1,_Neptunium_Series).svg.

In nature, during the time of the earth's formation, they have all withered away due to their very short lifespan and absence of a continuous supply (Bohr and Wheeler, 1939; Hilscher and Rossner, 1992). As a result, there are no fundamental differences between radioisotopes occurring in plants and those created by humans. To the four naturally occurring radioactive categories, the decay of manmade isotopes produces a slew of additional series. These series are known as collateral series and thus are comparable to natural ones (Specht, 1974; Schunck and Robledo, 2016).

5.6. ENERGY IN NUCLEAR REACTIONS

Take into account the nuclear reaction depicted in Figure 5.13 to categorize the nuclear reactions mentioned thus far. Preliminary reactants consist of a particulate x with mass mx having to move with velocity vx to a stationary specific component X with mass MX (Fraser and Milton, 1966; Krappe and Pomorski, 2012). Well after nuclear reaction, a substance with mass y and velocity v_y escapes whereas the product nucleus remains (Figure 5.14).

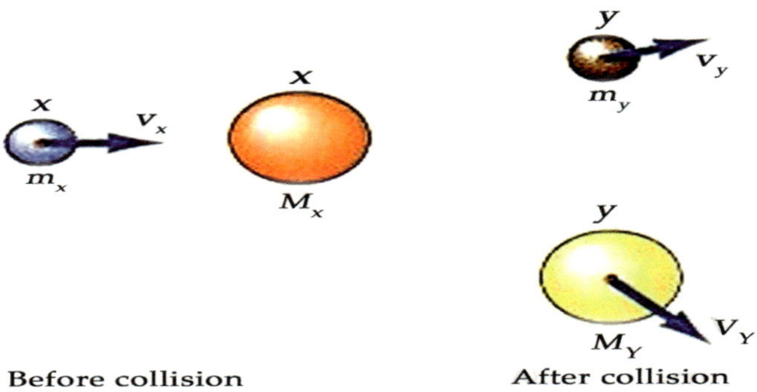

Figure 5.14. Nuclear reaction as a collision.

Source: https://byjus.com/chemistry/nuclear-reaction/.

The Y mass V_Y is the velocity of M_Y. The nuclear reaction can be written in a generic format as:

$$x + X = y + Y \qquad (30)$$

In which x and X represent the reagents and y and Y represent the reaction's products (Fong, 1956). Applying the principle of conserving energy to this reaction, we obtain:

$$mxc^2 + \mathrm{KE}x + MXc^2 = myc^2 + \mathrm{KE}y + MYc^2 + \mathrm{KE}Y$$

Rearranging,

$$(mx + MX)c^2 - (my + MY)c^2 = \mathrm{KE}y + \mathrm{KE}Y - \mathrm{KE}x$$

Currently, the Q factor of a nuclear reaction is described as such energy obtainable in a reaction due to the mass differential among the reacting substances.

Thus,

$$Q = (m_x + M_X)c^2 - (m_y - M_Y)c^2 \qquad (31)$$

or,

$$Q = ((\text{Input mass}) - (\text{Output mass}))c^2 \qquad (32)$$

or,

$$Q = E\text{in} - E\text{out} \tag{33}$$

Thus, the Q value is the differential between the energy input and the energy output of a nuclear reaction,

*E*in and the energy that comes out *E*out.

Q is larger than zero (Q > 0) if mx + MX is bigger than my + MY. In other words, the input mass energy exceeds the output mass-energy. Thus, the nuclear reaction results in the loss of mass and the release of energy Q. A nuclear event that releases energy is known as an exoergic reaction (often known as an exothermic reaction) (Bender et al., 2020).

If mx + MX is smaller as compared to my + MY in a nuclear reaction, therefore the Q value is negative (Q 0). In a process, mass is produced when Q joules of energy is introduced into the system. Typically, this energy is supplied through the kinetic energy (KE) of a responding molecule and nucleus. A nuclear reaction that adds energy to the system is known as an endoergic reaction (often known as an endothermic reaction) (Nix and Swiatecki, 1965; Gupta et al., 1975).

In terms of achieving lesser equilibrium energy levels, the nucleus produces a particle during the breakdown of a naturally occurring radioactive nuclide. Excessive energy is lost throughout the operation. In contrast, endoergic reactions do not happen spontaneously in the physical realm since the potential of the reagents is much less than the energy necessary to form the products. Therefore, endoergic responses cannot occur without the addition of energy to the network. The energy is supplied by continuing to accelerate the atom in an amplifier. This extra KE is required for the reaction to continue whenever the particle collides with its target. Perhaps more KE is required to pass the Coulomb barrier (Halpern, 1959; Nix, 1969).

It must be observed that the number of energies required to disassemble the nucleus, its Q number, is identical to the nucleus' BE as described before. Therefore, we may express a nuclear process using the form,

$$x + X \rightarrow y + Y + Q \tag{34}$$

When Q is more than zero, Q reflects the quantity of energy generated in a chemical process. When Q is less than zero, Q represents the quantity of energy that should be supplied to the system for the response to continue (Andreyev et al., 2017; Schmidt and Jurado, 2018).

5.7. NUCLEAR FISSION

Starting at the lower of the periodic table, Enrico Fermi shot neutrons at every chemical substance to generate isotopes of the components in 1934. He moved upwards to the periodic table till he reached uranium, the very last recognized element at the moment. He reasoned that the hitting uranium with neutrons will indeed destabilize it. He then reasoned that if the unsteady uranium nucleus underwent beta decay, the atomic number might well raise from 92 to 93, indicating that he had made a new component. (He is credited with coining the term transuranic.) He couldn't find what the byproducts were after the uranium bombing, though (Kuroda, 1960; Bulgac et al., 2020).

Otto Hahn and Lise Meitner reproduced the experimentations in Germany from 1935 to 1938. Ida Noddack, a German chemist, examined the byproducts and concluded that the uranium atom had already been broken into two neutral atoms. These findings led Lise Meitner and her nephew, Otto Frisch, to conclude that the atom had also been broken into two lighter elements. The dissociation of an atom closely resembles the division of a single life form into two equal-sized cells. Fission is the name for this biological mechanism. The dividing of an atom was then described by Otto Frisch using the biological word fission. As a result, nuclear fission is the division of a massive atom into two smaller elements (Balantekin and Takigawa, 1998; Schmidt et al., 2000). (_92235)U is the isotope of uranium that experiences fission. (_01) is an overall procedure,

$$ {}_{0}^{1}n + {}_{92}^{235}U \rightarrow \text{y+Y} + {}_{0}^{1}n + Q \qquad (35) $$

Nevertheless, nuclear fission may not always result in similar pieces. The result or component nuclei, y, and Y, were discovered to differ among the elements, Z ranges from 36 to 60. Common fission reactions include:

$$ {}_{0}^{1}n + {}_{92}^{235}U \rightarrow {}_{56}^{141}Ba + {}_{36}^{92}Kr + 3\,{}_{0}^{1}n + Q \qquad (36) $$

$$ {}_{0}^{1}n + {}_{92}^{235}U \rightarrow {}_{56}^{144}Ba + {}_{36}^{89}Kr + 3\,{}_{0}^{1}n + Q \qquad (37) $$

$$ {}_{0}^{1}n + {}_{92}^{235}U \rightarrow {}_{56}^{140}Xe + {}_{38}^{94}Sr + 2\,{}_{0}^{1}n + Q \qquad (38) $$

$$ {}_{0}^{1}n + {}_{92}^{235}U \rightarrow {}_{50}^{132}Sn + {}_{42}^{101}Mo + 3\,{}_{0}^{1}n + Q \qquad (39) $$

The weights of the resultant nuclei are always smaller than the weights of the reagents, showing a Q value larger than zero (Bradshaw et al., 2011). As a result, the reaction is exoergic, and energy is released in the procedure.

Figure 5.15 depicts the liquid-drop model, a conceptual model of nuclear fission created by Niels Bohr as well as John A. Wheeler.

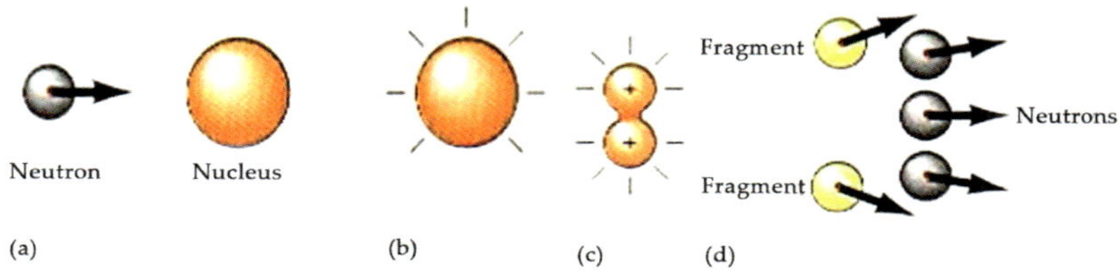

Figure 5.15. The liquid drops model of nuclear fission.

Source: https://commons.wikimedia.org/wiki/File:Liquid_drop_model_of_nuclear_fission.jpg.

When a focused beam neutron captures the uranium nucleus, it becomes irregular, oscillates, and deforms, as shown in Figure 5.15(c). Since the nucleus is expanded further apart in the temporary shape, the force of the nucleus is reduced. The Coulomb's repulsive force, on the other hand, remains as powerful as ever, splitting the drop (nucleus) into pieces (Figure 5.15(d)). As a result, the uranium nucleus splits into component nuclei with additional neutrons and a lot of energy (Winter, 2000).

In nuclear fission, about 90 distinct offspring nuclei are created. Slow neutrons are the first neutrons utilized to hit uranium as they contain extremely low KE and, as a result, low speeds and thus travel gradually. Since slow neutrons move so quietly, they get a high chance of being captured by the Uranium-235 nucleus. Every fission release around two or three neutrons (Rieth et al., 2013).

It could be useful to include a brief story on nuclear fission below. "If it were ever discovered practicable to regulate at will the speed of breakdown of the radioactive materials, a vast quantity of energy might be gained from a slight quantity of matter," Lord Rutherford declared in 1906 at McGill University in Montreal, Canada. 1 "The energy created by collapsing the atom is a pretty poor type of stuff," Rutherford would later say as he grew older. Everyone who believes that the transition of such atoms will be a power source is speaking moonshine." That comment was a question to Rutherford's Hungarian scientist Leo Szilard (1898–1964). "Imagine if you discovered an atom wherein the nuclei emit energy?" Szilard wondered. Imagine if you can somehow make things work whenever you wanted? What if the atoms of this element emitted two more neutrons, striking two extra nuclei? Two twos equal four, four fours equal 16 — the total would be enormous in a flash. Moonshine? All you have to do now is locate the appropriate ingredient!

Fission creates the same particles that started the fission in the first place, notably neutrons, as a by-product. A multiplication occurs when more neutrons are created than when the reaction began. Additional neutrons are created to produce more fission if the extra product neutrons may ignite more fission, and so on. As seen in Figure 5.16, the consequence is a chain reaction (Jones et al., 1989; Linke et al., 2019).

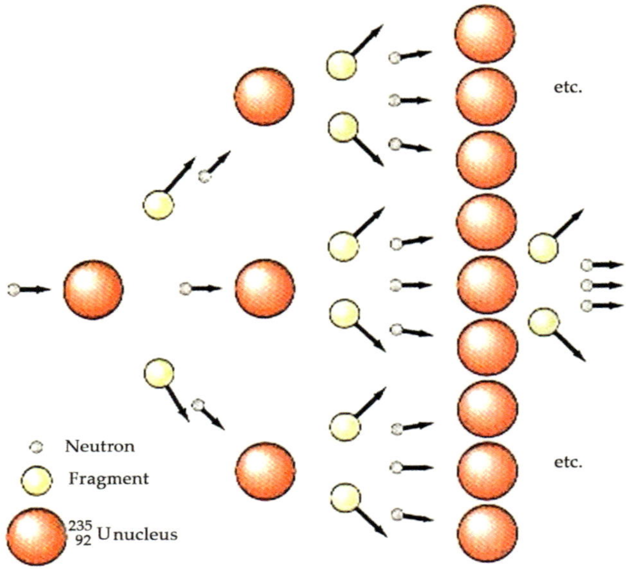

Figure 5.16. The chain reaction.

Source: https://www.atomicarchive.com/science/fission/chain-reactions.html.

A multiplication factor, k, is used to multiply neutrons. If k 1, the reaction produces fewer neutrons than it did at the start, as well as the chain, breaks down. If $k > 1$, the process produces too many neutrons and spirals out of control. If $k = 1$, the exact number of neutrons expected to maintain the operation running at a continuous rate is created (Ichimaru, 1993).

Natural uranium is 99.3% (_92238)U and just 0.7% (_92235)U, and it is incapable of chain reaction. The proportion of (_92235)U must be raised to achieve a chain reaction.

Nuclear reactor level uranium has only approximately 3.6% of (_92235) U, which is much too tiny to cause a nuclear explosion. Weapons-grade uranium has around 50% of (_92235)U. To wrap up our brief narrative, on June 28, 1934, Leo Szilard submitted a patent application with the London Patents Office. It was the globe's first neutron bombardment recording of a nuclear processing chain of events. Szilard was concerned that his chain reaction concept would end up in the hands of the Nazis, so he appointed his invention to the British Admiralty. Overall, the more nuclei that are divided, the more energy is released. Szilard made the error of suggesting splitting the alkali metals rather than the heavier ones (McClintock, 1942).

5.7.1. The Atomic Bomb

A few nuclear scientists conceived of creating an atomic weapon due to the potential of a chain of events in uranium, which would result in a tremendous energy release. The Second World War was ranting and raving in Europe, and researchers feared Hitler could create a real bomb. It was believed that this bomb in his hands would lead to the demise of civilization. The Hungarian scientists who'd already fled Hitler's Europe, Leo Szilard, and Edward Teller (soon becoming the "Dad of the Hydrogen Weapon"), contacted Albert Einstein and asked him to prepare a statement to President Roosevelt about the possibilities of creating an atomic bomb. On October 11, 1939, the message was directly presented to President Roosevelt by Dr. Alexander Sachs. Strangely, the ultimate proposal for the construction of the A-bomb was taken under the name Manhattan Project on December 6, 1941. To create an atomic weapon, sufficient Uranium-235 was required to initiate the chain reaction. The quantity of Uranium-235 required to initiate a chain reaction is known as the critical mass (Miyamoto, 2005).

The Uranium-235 was required to be comprised of two subcritical pieces. When one component, in the shape of a bullet, was shot into the 2nd part, the saturation point was reached and an explosion would result from the chain of events. This was the kind of bomb known as "Thin Man" that exploded on August 5, 1945, at Hiroshima. It was hard to isolate (_92235)U from (_92238)U, which posed a challenge for uranium bombs.

Neutron bombardment of (_92238)U yields the isotope neptunium, (_93239)Np, that decays into plutonium, (_94239)Pu, as previously observed in Eqn. (28). It seems that Plutonium-239 is more fissile material than Uranium-235, requiring a far less amount for its critical mass. Through the construction of a nuclear reactor, as we shall explore in a minute, a vast and very inexpensive source of plutonium became accessible. Therefore, the Manhattan Project continued to create a plutonium weapon, a different sort of atomic bomb. As seen in Figure 5.17, the plutonium bomb was constructed in the shape of a spherical with bits of plutonium, including one with a mass underneath the saturation point, and along the perimeter. For igniting, a succession of chemical blasts simultaneously propelled the plutonium fragments into the sphere's core. While all of such plutonium fragments got together, the saturation point of plutonium was reached, the chain of events was triggered, and the bomb detonated (Ongena and Ogawa, 2016). The initial test of an atomic weapon was conducted on July 16, 1945, at a location known as "Trinity" in the New Mexico desert using a plutonium bomb. On August 9, 1945, the very first plutonium bomb, dubbed "Fat Boy," was shot down on Nagasaki.

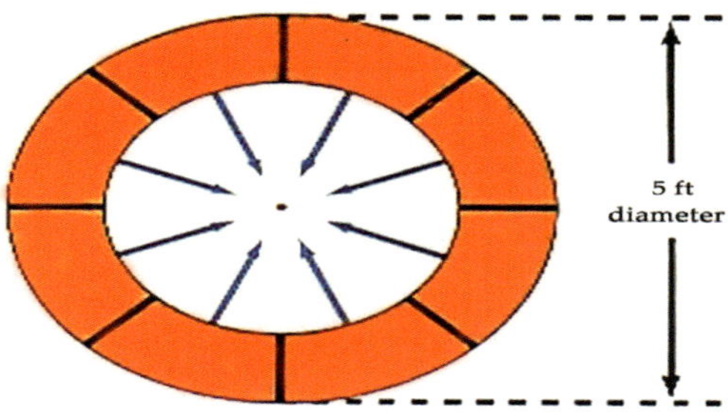

Figure 5.17. Triggering the plutonium bomb.

Source: https://www.britannica.com/technology/nuclear-weapon/Gun-assembly-implosion-and-boosting.

5.7.2. Fission Nuclear Reactors

Enrico Fermi developed the very first nuclear reactor on the volleyball court beneath the west seats of Stagg Field at the University of Chicago. It commenced operations on December 2, 1942, after being established in October 1942. It's the first-time nuclear fission was used in a controllable environment (Petrescu et al., 2017).

Figure 5.18 depicts the basic nuclear reactor. Uranium (_92238)U, enhanced with 3.6% (_92235)U, is used in the reactor. A reaction like Eqn. (36) produces neutrons. The neutrons emitted have a relatively sufficient KE and thus are known as fast neutrons due to their rapid speed and enormous KE. These rapid neutrons are traveling too quickly to start new fission reactions and should be decelerated. One method is to submerge the whole unit in a high-pressure water pool. A high-pressure water reactor is this one reactor (PWR). The neutrons now clash with water molecules and decelerate, allowing them to participate in nuclear fission. The neutrons are moderated or slowed by the water, which is termed the moderator. The slow neutrons will now divide more (_92235)U nuclei till a chain reaction occurs. The chain reaction also isn't permitted to go amok as with an atomic weapon but is instead regulated by some cadmium poles placed into the reactor (Conde and Fink, 1976).

Cadmium is a radioactive material that can collect a considerable amount of neutrons rather than become unstable. As a result, when the cadmium graphite rods are placed into the reactor, they capture neutrons, reducing the number of neutrons attainable for fission. The fission reaction may be regulated in this manner. The water moderator is also a cooling agent. Water is heated by the immense heat created by nuclear fission, which is then transported to a heat exchanger. The boiling point of water rises with pressure, and the hot modulator water is at such a high pressure and temperature. As a result, the moderator water may reach temperatures of several 100°C with no boiling (Clark, 2005). The secondary water cooling is heated whenever this water reaches the heating element With the High heat of the main coolant, the comparatively normal-pressure secondary coolant is rapidly turned to steam.

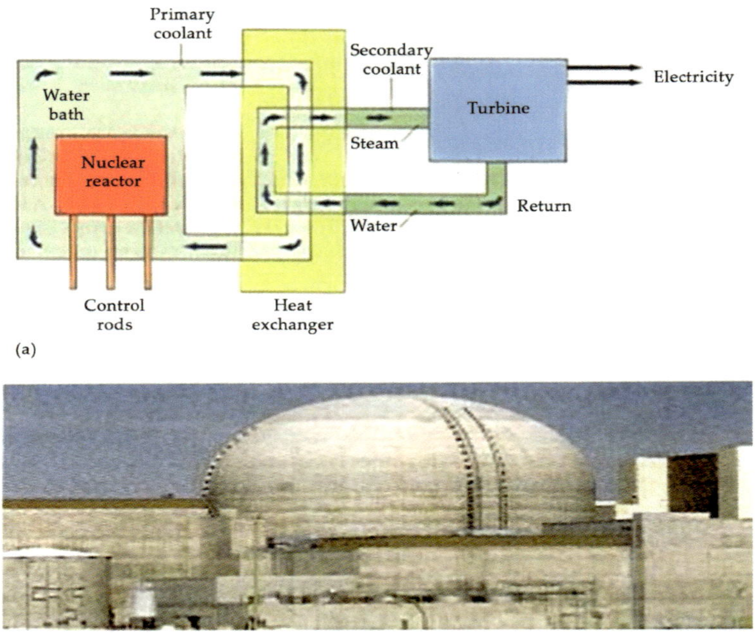

Figure 5.18. A typical nuclear reactor.

Source: https://world-nuclear.org/information-library/nuclear-fuel-cycle/nuclear-power-reactors/nuclear-power-reactors.aspx.

This steam is then transported via a turbine, which powers an electric generator, resulting in the generation of energy. A reactor produces a significant amount of energy. In the breaking of only one (_92235)U nucleus, roughly 200 MeV of energy is released. Chemical processes often produce energy on the order of 3 to 4 eV. As a result, fission of (_92235)U produces around 2.5 million different ways that much energy by combining a relatively similar quantity of carbon (like in coal or gasoline) (Lee et al., 1988).

The nuclear waste element is the only disadvantage of a fission reactor. Fission pieces such as (_56141) are represented in Eqns. (36) to (39). $^{141}_{56}Ba$, $^{92}_{36}Kr$, $^{144}_{56}Ba$, $^{89}_{36}Kr$, $^{140}_{54}Xe$, $^{94}_{38}Sr$, $^{132}_{50}Sn$, $^{101}_{42}Mo$ Some of the reaction's probable results are Mo. These are unsteady isotopes that disintegrate into other radioactive nuclei. All of these deadly radioactive waste nuclei should ultimately be disposed of. Because a few of them have large half-lives, they will be here for a long period. They can't be discarded in the seas or left in any other location where they would pollute the environment, like the air or soil (Toschi, 1997).

They should not be permitted to consume the water. The greatest area to keep these wastes has so far been discovered in the bottoms of historic salt mines, which are extremely dry and thousands of feet underneath the earth's surface. They may decompose with avoiding damaging the atmosphere here. Many individuals have an irrational worry that a nuclear reactor would burst like just an atomic bomb and destroy everyone in its vicinity. That there is not sufficient (92235)U in a nuclear reactor to make an atomic weapon. Furthermore, the natural location of the cadmium rod is inside the reactor. They should be removed to start and maintain the reactor. The graphite rods simply rely upon the reactors whether any of the reactor's mechanisms fail, interrupting the chain reaction as well as powering down the reactor.

The breeder reactor is the other form of fission nuclear reactor. A breeder reactor employs uranium (_92238)U or thorium (_90232)Th as nuclear fuel rather than the slow neutrons employed in a PWR. As per Eqn. (28), fast neutrons interact to (_92238)U to generate neptunium, (_93239)Np.

As per Eqn. (29), neptunium (_93239)Np decays to generate plutonium (_94239)Pu. Because plutonium is extremely spontaneous fission, it may also provide power to the reactor. The formation of plutonium in the reactor produces more fissile material as compared to what is consumed. The term breeder reactor comes from the fact that its "breeds" nuclear fuel. The breeder reactor may, of course, produce power while producing additional fuel. To make plutonium for nuclear weapons, breeder reactors are employed (Dasgupta et al., 2007).

5.8. NUCLEAR FUSION

It has historically been recognized that the sun generates lots of energy for a vast amount of time. The origin of this power was the subject of great discussion. Hans Bethe (1906–2005) proposed in 1938 that the huge energy generated was due to the fusing of hydrogen nuclei into helium atoms nuclear fusion is a reaction wherein lighter atoms combine to form a heavier atom with a lot of energy. The energy was produced in the sun, according to Bethe, via the proton-proton process. Two protons combine to generate an unsteady isotope of helium in the initial stage of the cycle (Aoki et al., 2012).

$^{1}_{1}\mathrm{p} + ^{1}_{1}\mathrm{p} \rightarrow ^{2}_{2}\mathrm{He}$

However, a few of such coupled protons in the unsteady isotope's nucleus decays instantly by Eqn. (20) as:

$$^{1}_{1}p \rightarrow ^{1}_{0}n + ^{0}_{+1}e + \mathrm{v}$$

The neutron now joins the initial proton to produce the deuteron, and the process may be written as,

$$^{1}_{1}p + ^{1}_{1}p \rightarrow ^{2}_{1}H + ^{0}_{+1}e + \mathrm{v} \qquad (40)$$

The weaker nuclear force causes the disintegration of a proton or neutron in the nucleus, which we shall discuss in further depth in the next chapters. Per the process, the deuteron generated in Eqn. (40) now unites with some other protons to create the helium isotope $^{3}_{2}He$,

$$^{2}_{1}H + ^{1}_{1}p \rightarrow ^{3}_{2}He + \gamma \qquad (41)$$

Eqns. (40) and (41) should be repeated twice to produce two (_23)He nuclei, that either interact as per Eqn. (42),

$$^{3}_{2}He + ^{3}_{2}He \rightarrow ^{4}_{2}He + 2^{1}_{1}p \qquad (42)$$

As a result of the fusing of the nuclei of the hydrogen atom, the permanent element helium was created. The full proton-proton cycle may be written in abbreviation as,

$$2^{1}_{1}p \rightarrow ^{4}_{2}He + 2^{0}_{+1}e + 2y + 2v + Q \qquad (43)$$

The system's total Q value, or power produced, is around 26 MeV.

5.8.1. The Hydrogen Bomb

In July 1942, Robert Oppenheimer (1904–1967), writing on the research of Edward Teller, Enrico Fermi, and Hans Bethe, remarked that the extraordinarily high heat of an atomic bomb may have been consumed to ignite a nuclear fusion in deuterium, resulting in a fusion bomb or a hydrogen bomb. The reaction of deuterium and tritium, all hydrogen isotopes, Deuterium is quite prevalent in ocean water, but tritium is limited. Tritium, on the other hand, may be produced in a nuclear reactor by encircling the core with lithium (Botta et al., 2012).

As a result, all of the tritium required may be produced quite simply. As seen in Figure 5.19, the hydrogen bomb is essentially a bomb inside a bomb. A typical plutonium-based atomic weapon is detonated (Klein and Marshalek, 1991). The huge heat generated by the A-bomb provides the high temperature required to initiate the fusion of the deuterium-tritium combination. The critical mass of plutonium reduces the range of an A-bomb. We are unable to manufacture a sufficient quantity of plutonium higher than that of the critical mass with no causing it to explode.

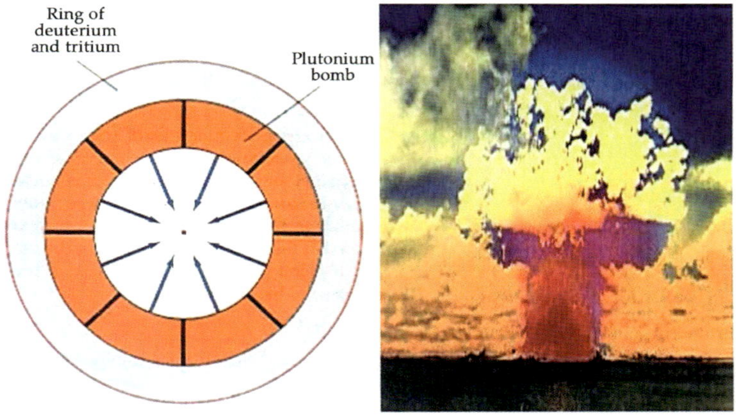

Figure 5.19. A hydrogen bomb. The energy released in an explosion resulting.

Source: https://www.britannica.com/technology/nuclear-weapon.

Nevertheless, we can make quite so many deuterium and tritium as we would like. That will never go off so that fed the extraordinarily high temperature required for fusion (Cohen et al., 1995). On October 31, 1952, the first H-bomb exploded. It destroyed the Marshall Islands Island of Eniwetok. On August 12, 1953, the Soviet Union joined forces by blowing their H-Bomb. Since lithium is less expensive and thus more readily accessible, the Soviets used it for tritium in the fusion process (Beane et al., 2011).

5.8.2. The Fusion Reactor

One of the challenges of a fission reactor is the nuclear waste or pieces that result from the process. The fusing technique yields no radioactive by-products. Such that, the sole product of fusion is helium, which is a non-radioactive inert gas. The sun's proton-proton cycle is too slow to be replicated in a reactor. As a result, a fusion reactor's fusion cycle (Woosley et al., 2004).

The incredibly high temperatures connected with the fusion reaction, that is, trillions of kelvins, make designing a fusion reactor problematic. (Note that the sun's surface temperature is around 6,000 K, and its central temperature is millions of times greater.) All materials capable of containing the process would burn at such temperatures. Nevertheless, developing a fusion reactor is not unachievable, although it is complex. Electrons and nuclei are disconnected out of one other in plasma, a charged fluid, just at high temperatures of fusion. The fusion process may be confined inside magnetic fields due to the fluid's electric charges. A magnetic restriction has been used to build innovative nuclear fusion on a small scale, with certain success (Käppeler et al., 2011). The fusion reactor even now requires a huge amount of improvement. This phase is executed since this fusion reactor has the potential to generate huge amounts of energy from an inexpensive fuel, essentially water, with really no radioactive substances as a side product (Bonatsos and Daskaloyannis, 1999; Mitchell et al., 2010).

REFERENCES

1. Adloff, J. P., (1981). Hot atom chemistry in the natural radioactive series. *Radiochimica Acta, 29*(1), 5–8.
2. Andrews, D. G. H., (2009). An excel™ model of a radioactive series. *Physics Education, 44*(1), 48.
3. Andreyev, A. N., Nishio, K., & Schmidt, K. H., (2017). Nuclear fission: A review of experimental advances and phenomenology. *Reports on Progress in Physics, 81*(1), 016301.
4. Aoki, S., Doi, T., Hatsuda, T., Ikeda, Y., Inoue, T., Ishii, N., & (HAL QCD Collaboration), (2012). Lattice quantum chromodynamical approach to nuclear physics. *Progress of Theoretical and Experimental Physics, 2012*(1), 01A105.
5. Arthur, M. D., Brysk, H., Paveri-Fontana, S. L., & Zweifel, P. F., (1981). The law of radioactive decay. *Il Nuovo Cimento B (1971–1996), 63*(2), 565–587.
6. Aston, P. J., (2012). Is radioactive decay really exponential?. *EPL (Europhysics Letters), 97*(5), 52001.
7. Bakaç, M., Taşoğlu, A. K., & Uyumaz, G., (2011). Modeling radioactive decay. *Procedia-Social and Behavioral Sciences, 15*, 2196–2200.
8. Balantekin, A. B., & Takigawa, N., (1998). Quantum tunneling in nuclear fusion. *Reviews of Modern Physics, 70*(1), 77.
9. Baldin, A. M., (1980). *Relativistic Nuclear Physics (No. JINR-E—1-80-174)*. Joint Inst. for Nuclear Research.
10. Beane, S. R., Detmold, W., Orginos, K., & Savage, M. J., (2011). Nuclear physics from lattice QCD. *Progress in Particle and Nuclear Physics, 66*(1), 1–40.
11. Bender, M., Bernard, R., Bertsch, G., Chiba, S., Dobaczewski, J., Dubray, N., & Åberg, S., (2020). Future of nuclear fission theory. *Journal of Physics G: Nuclear and Particle Physics, 47*(11), 113002.
12. Bergman, O., Lippert, M., & Lifschytz, G., (2007). Holographic nuclear physics. *Journal of High Energy Physics, 2007*(11), 056.
13. Bethe, H. A., & Bacher, R. F., (1936). Nuclear physics A. Stationary states of nuclei. *Reviews of Modern Physics, 8*(2), 82.
14. Bethe, H. A., (1937). Nuclear physics B. Nuclear dynamics, theoretical. *Reviews of Modern Physics, 9*(2), 69.

15. Blaum, K., Dilling, J., & Nörtershäuser, W., (2013). Precision atomic physics techniques for nuclear physics with radioactive beams. *Physica Scripta, 2013*(T152), 014017.
16. Bohigas, O., & Weidenmuller, H. A., (1988). Aspects of chaos in nuclear physics. *Annual Review of Nuclear and Particle Science, 38*(1), 421–453.
17. Bohr, A. N., & Mottelson, B. R., (1998). *Nuclear Structure (In 2 Volumes)*. World Scientific Publishing Company.
18. Bohr, N., & Wheeler, J. A., (1939). The mechanism of nuclear fission. *Physical Review, 56*(5), 426.
19. Bonatsos, D., & Daskaloyannis, C., (1999). Quantum groups and their applications in nuclear physics. *Progress in Particle and Nuclear Physics, 43*, 537–618.
20. Botta, E., Bressani, T., & Garbarino, G., (2012). Strangeness nuclear physics: A critical review on selected topics. *The European Physical Journal A, 48*(3), 1–64.
21. Bozkurt, A. H. M. E. T., Yorulmaz, N., & Kam, E., (2007). Environmental radioactivity measurements in Harran plain of Sanliurfa, Turkey. In: *AIP Conference Proceedings* (Vol. 899, No. 1, pp. 397–398). American Institute of Physics.
22. Bradshaw, A. M., Hamacher, T., & Fischer, U., (2011). Is nuclear fusion a sustainable energy form?. *Fusion Engineering and Design, 86*(9–11), 2770–2773.
23. Bretscher, E., (1940). Radioactivity of Be10. *Nature, 146*(3690), 94, 95.
24. Brown, G. E., (1987). Relativistic effects in nuclear physics. *Physica Scripta, 36*(2), 209.
25. Bulgac, A., Jin, S., & Stetcu, I., (2020). Nuclear fission dynamics: Past, present, needs, and future. *Frontiers in Physics, 8*, 63.
26. Butt, D. K., & Wilson, A. R., (1972). A study of the radioactive decay law. *Journal of Physics A: General Physics, 5*(8), 1248.
27. Carlson, J., Gandolfi, S., Pederiva, F., Pieper, S. C., Schiavilla, R., Schmidt, K. E., & Wiringa, R. B., (2015). Quantum Monte Carlo methods for nuclear physics. *Reviews of Modern Physics, 87*(3), 1067.
28. Casten, R., & Casten, R. F., (2000). *Nuclear Structure from a Simple Perspective* (Vol. 23). Oxford University Press on Demand.

29. Clark, R. E., (2005). *Nuclear Fusion Research: Understanding Plasma-Surface Interactions* (Vol. 78). Springer Science & Business Media.
30. Coen, D. R., (2002). Scientists' errors, nature's fluctuations, and the law of radioactive decay, 1899–1926. *Historical Studies in the Physical and Biological Sciences, 32*(2), 179–205.
31. Cohen, T. D., Furnstahl, R. J., Griegel, D. K., & Jin, X., (1995). QCD sum rules and applications to nuclear physics. *Progress in Particle and Nuclear Physics, 35*, 221–298.
32. Conde, J., & Fink, G. R., (1976). A mutant of Saccharomyces cerevisiae defective for nuclear fusion. *Proceedings of the National Academy of Sciences, 73*(10), 3651–3655.
33. Cooper, P. S., (2009). Searching for modifications to the exponential radioactive decay law with the Cassini spacecraft. *Astroparticle Physics, 31*(4), 267–269.
34. Corral, A., Font, F., & Camacho, J., (2011). Noncharacteristic half-lives in radioactive decay. *Physical Review E, 83*(6), 066103.
35. Cruz-López, C. A., & Espinosa-Paredes, G., (2022). Fractional radioactive decay law and Bateman equations. *Nuclear Engineering and Technology, 54*(1), 275–282.
36. Dasgupta, M., Hinde, D. J., Diaz-Torres, A., Bouriquet, B., Low, C. I., Milburn, G. J., & Newton, J. O., (2007). Beyond the coherent coupled channels description of nuclear fusion. *Physical Review Letters, 99*(19), 192701.
37. Davis, J. J., Coopey, R. W., Watson, D. G., Palmiter, C. C., & Cooper, C. L., (1951). The radioactivity and ecology of aquatic organisms in the Columbia River. *Biology Research – Annual Report.*
38. De Marcillac, P., Coron, N., Dambier, G., Leblanc, J., & Moalic, J. P., (2003). Experimental detection of α-particles from the radioactive decay of natural bismuth. *Nature, 422*(6934), 876–878.
39. Doğru, M., & Külahci, F., (2004). Iso-radioactivity curves of the water of the Hazar Lake, Elazig, Turkey. *Journal of Radioanalytical and Nuclear Chemistry, 260*(3), 557–562.
40. Durante, M., & Paganetti, H., (2016). Nuclear physics in particle therapy: A review. *Reports on Progress in Physics, 79*(9), 096702.
41. Elliott, D. R., & King, L. D. P., (1941). Extension of the radioactive series, Z= N±1. *Physical Review, 60*(7), 489.

42. Engel, J., Pittel, S., & Vogel, P., (1992). Nuclear physics of dark matter detection. *International Journal of Modern Physics E, 1*(01), 1–37.
43. English, A. C., Cranshaw, T. E., Demers, P., Harvey, J. A., Hincks, E. P., Jelley, J. V., & May, A. N., (1947). The (4 n+ 1) radioactive series. *Physical Review, 72*(3), 253.
44. Fadel, K., (2014). Diagnosis, care, and healing. Radioactivity serves medicine. *Decouverte (Paris)*, 22–31.
45. Fajans, K., (2013). Radioactive transformations and the periodic system of the elements. In: *A Source Book in Chemistry, 1900–1950* (pp. 67–75). Harvard University Press.
46. Feshbach, H., (1958). Unified theory of nuclear reactions. *Annals of Physics, 5*(4), 357–390.
47. Fong, P., (1956). Statistical theory of nuclear fission: Asymmetric fission. *Physical Review, 102*(2), 434.
48. Frankfurt, L., & Strikman, M., (1988). Hard nuclear processes and microscopic nuclear structure. *Physics Reports, 160*(5, 6), 235–427.
49. Fraser, J. S., & Milton, J. C. D., (1966). Nuclear fission. *Annual Review of Nuclear Science, 16*(1), 379–444.
50. Gal, A., Hungerford, E. V., & Millener, D. J., (2016). Strangeness in nuclear physics. *Reviews of Modern Physics, 88*(3), 035004.
51. Godovikov, S. K., (2004). Nonexponential 125mTe radioactive decay. *Journal of Experimental and Theoretical Physics Letters, 79*(5), 196–199.
52. Gopych, P. M., & Zalyubovskii, I. I., (1988). Is the basic law of radioactive decay exponential. *Sov. J. Particles Nucl.(Engl. Transl.);(United States), 19*(4).
53. Grawe, H., Langanke, K., & Martínez-Pinedo, G., (2007). Nuclear structure and astrophysics. *Reports on Progress in Physics, 70*(9), 1525.
54. Grosse, A. V., Booth, E. T., & Dunning, J. R., (1941). The fourth (4 n+ 1) radioactive series. *Physical Review, 59*(3), 322.
55. Gupta, R. K., Scheid, W., & Greiner, W., (1975). Theory of charge dispersion in nuclear fission. *Physical Review Letters, 35*(6), 353.
56. Hagemann, F., Katzin, L. I., Studier, M. H., Ghiorso, A., & Seaborg, G. T., (1947). *The 4n+ 1 Radioactive Series: The Decay Products of U233* (Vol. 1186). Technical information division, Oak Ridge Directed Operations.

57. Hagemann, F., Katzin, L. I., Studier, M. H., Seaborg, G. T., & Ghiorso, A., (1950). The 4n+ 1 radioactive series: The decay products of U 233. *Physical Review, 79*(3), 435.

58. Halpern, I., (1959). Nuclear fission. *Annual Review of Nuclear Science, 9*(1), 245–342.

59. Haxel, O., Jensen, J. H. D., & Suess, H. E., (1949). On the "magic numbers" in nuclear structure. *Physical Review, 75*(11), 1766.

60. Henderson, M. C., Livingston, M. S., & Lawrence, E. O., (1934). Artificial radioactivity produced by deuton bombardment. *Physical Review, 45*(6), 428.

61. Hilscher, D., & Rossner, H., (1992). Dynamics of nuclear fission. In *Annales de Physique* (Vol. 17, No. 6, pp. 471–552). EDP Sciences.

62. Hodgson, P. E., (1971). *Nuclear Reactions and Nuclear Structure* (Vol. 426). Oxford: Clarendon Press.

63. Hofstadter, R., (1956). Electron scattering and nuclear structure. *Reviews of Modern Physics, 28*(3), 214.

64. Huestis, S. P., (2002). Understanding the origin and meaning of the radioactive decay equation. *Journal of Geoscience Education, 50*(5), 524–527.

65. Ichimaru, S., (1993). Nuclear fusion in dense plasmas. *Reviews of Modern Physics, 65*(2), 255.

66. Jesse, K. E., (2003). Computer simulation of radioactive decay. *The Physics Teacher, 41*(9), 542, 543.

67. Jones, S. E., Palmer, E. P., Czirr, J. B., Decker, D. L., Jensen, G. L., Thorne, J. M., & Rafelski, J., (1989). Observation of cold nuclear fusion in condensed matter. *Nature, 338*(6218), 737–740.

68. Käppeler, F., Gallino, R., Bisterzo, S., & Aoki, W., (2011). The s process: Nuclear physics, stellar models, and observations. *Reviews of Modern Physics, 83*(1), 157.

69. Kielland-Brandt, M. C., (1974). Studies on the biosynthesis of tobacco mosaic virus: VII. Radioactivity of plus and minus strands in different forms of viral RNA after labelling of infected tobacco leaves. *Journal of Molecular Biology, 87*(3), 489–503.

70. Kirby, H. W., (1954). Decay and growth tables for naturally occurring radioactive series. *Analytical Chemistry, 26*(6), 1063–1071.

71. Klein, A., & Marshalek, E. R., (1991). Boson realizations of lie algebras with applications to nuclear physics. *Reviews of Modern Physics, 63*(2), 375.
72. Kondev, F. G., Wang, M., Huang, W. J., Naimi, S., & Audi, G., (2021). The NUBASE2020 evaluation of nuclear physics properties. *Chinese Physics C, 45*(3), 030001.
73. Koning, A. J., & Rochman, D., (2008). Towards sustainable nuclear energy: Putting nuclear physics to work. *Annals of Nuclear Energy, 35*(11), 2024–2030.
74. Konopinski, E. J., & Uhlenbeck, G. E., (1941). On the Fermi theory of β-radioactivity. II. The "forbidden" spectra. *Physical Review, 60*(4), 308.
75. Krappe, H. J., & Pomorski, K., (2012). *Theory of Nuclear Fission: A Textbook* (Vol. 838). Springer Science & Business Media.
76. Krause, D. E., Rogers, B. A., Fischbach, E., Buncher, J. B., Ging, A., Jenkins, J. H., & Sturrock, P. A., (2012). Searches for solar-influenced radioactive decay anomalies using spacecraft RTGs. *Astroparticle Physics, 36*(1), 51–56.
77. Kulagin, S. A., & Petti, R., (2006). Global study of nuclear structure functions. *Nuclear Physics A, 765*(1, 2), 126–187.
78. Kuroda, P. K., (1960). Nuclear fission in the early history of the earth. *Nature, 187.*
79. Langevin, M., (1985). *New Forms of Radioactivity (No. LYCEN--8502) (pp. 1-25).*
80. Lee, B., Kim, Y., L'yi, W., Kim, J., Seo, B., & Hong, S., (2021). Radiological analysis for radioactivity depth distribution in activated concrete using gamma-ray spectrometry. *Applied Radiation and Isotopes, 169*, 109558.
81. Lee, S., Tou, T. Y., Moo, S. P., Eissa, M. A., Gholap, A. V., Kwek, K. H., & Zakaullah, M., (1988). A simple facility for the teaching of plasma dynamics and plasma nuclear fusion. *American Journal of Physics, 56*(1), 62–68.
82. Linke, J., Du, J., Loewenhoff, T., Pintsuk, G., Spilker, B., Steudel, I., & Wirtz, M., (2019). Challenges for plasma-facing components in nuclear fusion. *Matter and Radiation at Extremes, 4*(5), 056201.
83. Livingston, M. S., & Bethe, H. A., (1937). Nuclear physics C. nuclear dynamics, experimental. *Reviews of Modern Physics, 9*(3), 245.

84. McClintock, B., (1942). The fusion of broken ends of chromosomes following nuclear fusion. *Proceedings of the National Academy of Sciences of the United States of America, 28*(11), 458.

85. Measday, D. F., (2001). The nuclear physics of muon capture. *Physics Reports, 354*(4, 5), 243–409.

86. Meißner, U. G., (2015). Anthropic considerations in nuclear physics. *Science Bulletin, 60*(1), 43–54.

87. Miller, J. M., & Hudis, J., (1959). High-energy nuclear reactions. *Annual Review of Nuclear Science, 9*(1), 159–202.

88. Mitchell, G. E., Richter, A., & Weidenmüller, H. A., (2010). Random matrices and chaos in nuclear physics: Nuclear reactions. *Reviews of Modern Physics, 82*(4), 2845.

89. Miyamoto, K., (2005). *Plasma Physics and Controlled Nuclear Fusion* (Vol. 38). Springer Science & Business Media.

90. Mobaligh, M., Talbi, A., Alahyane, A., El Fakir, M. B., Misdaq, M. A., & Fares, K., (2021). Assessment of composts quality produced from phosphate washing sludge and leachates, based on radioactivity, phytotoxicity and different phosphorus forms. *Plant Cell Biotechnology and Molecular Biology,* 595–609.

91. Nifenecker, H., & Pinston, J. A., (1990). High energy photon production in nuclear reactions. *Annual Review of Nuclear and Particle Science, 40*(1), 113–144.

92. Nix, J. R., & Swiatecki, W. J., (1965). Studies in the liquid-drop theory of nuclear fission. *Nuclear Physics, 71*(1), 1–94.

93. Nix, J. R., (1969). Further studies in the liquid-drop theory on nuclear fission. *Nuclear Physics A, 130*(2), 241–292.

94. Ohlsen, G. G., (1972). Polarization transfer and spin correlation experiments in nuclear physics. *Reports on Progress in Physics, 35*(2), 717.

95. Ongena, J., & Ogawa, Y., (2016). Nuclear fusion: Status report and future prospects. *Energy Policy, 96*, 770–778.

96. Papadopoulos, A., Christofides, G., Koroneos, A., Stoulos, S., & Papastefanou, C., (2013). Radioactive secular equilibrium in 238U and 232Th series in granitoids from Greece. *Applied Radiation and Isotopes, 75*, 95–104.

97. Patra, A. C., Mohapatra, S., Sahoo, S. K., Tripathi, R. M., & Puranik, V. D., (2013). Radioactive series disequilibria in underground uranium

deposits of the Singhbhum Shear Zone, Eastern India. *Journal of Radioanalytical and Nuclear Chemistry, 295*(1), 675–683.

98. Petrescu, R. V., Aversa, R., Kozaitis, S., Apicella, A., & Petrescu, F. I., (2017). Some proposed solutions to achieve nuclear fusion. *American Journal of Engineering and Applied Sciences, 10*(3).
99. Pfützner, M., (2013). Particle radioactivity of exotic nuclei. *Physica Scripta, 2013*(T152), 014014.
100. Pommé, S., & Pelczar, K., (2020). On the recent claim of correlation between radioactive decay rates and space weather. *The European Physical Journal C, 80*(11), 1–8.
101. Pommé, S., & Pelczar, K., (2022). Role of ambient humidity underestimated in research on correlation between radioactive decay rates and space weather. *Scientific Reports, 12*(1), 1–4.
102. Qi, C., Xu, F. R., Liotta, R. J., & Wyss, R., (2009). Universal decay law in charged-particle emission and exotic cluster radioactivity. *Physical Review Letters, 103*(7), 072501.
103. Radenković, M. B., Joksić, J. D., & Kovačević, J., (2015). Natural radionuclides content and radioactive series disequilibrium in drinking waters from Balkans region. *Journal of Radioanalytical and Nuclear Chemistry, 306*(1), 295–299.
104. Ribeiro, F. B., (1998). Simultaneous diffusion of isotopes from a radioactive series is homogeneous and isotropic solids. *Radiation Measurements, 29*(1), 9–18.
105. Rieppo, R., (1978). A method to investigate the secular equilibrium of natural radioactive series. *Geoexploration, 16*(3), 177–183.
106. Rieth, M., Dudarev, S. L., De Vicente, S. G., Aktaa, J., Ahlgren, T., Antusch, S., & Zivelonghi, A., (2013). Recent progress in research on tungsten materials for nuclear fusion applications in Europe. *Journal of Nuclear Materials, 432*(1–3), 482–500.
107. SAKANOUE, M., (1967). Age determinations by utilizing the disequilibrium systems of uranium and actinouranium radioactive series. *The Quaternary Research (Daiyonki-Kenkyu), 6*(4), 121–133.
108. Sargsian, M. M., (2001). Selected topics in high energy semi-exclusive electro-nuclear reactions. *International Journal of Modern Physics E, 10*(06), 405–457.
109. Sato, H., & Yazaki, K., (1981). On the coalescence model for high energy nuclear reactions. *Physics Letters B, 98*(3), 153–157.

110. Schmidt, B., Selmer, T., Ingendoh, A., & Von, F. K., (1995). A novel amino acid modification in sulfatases that is defective in multiple sulfatase deficiency. *Cell, 82*(2), 271–278.

111. Schmidt, K. H., & Jurado, B., (2018). Review on the progress in nuclear fission—Experimental methods and theoretical descriptions. *Reports on Progress in Physics, 81*(10), 106301.

112. Schmidt, K. H., Steinhäuser, S., Böckstiegel, C., Grewe, A., Heinz, A., Junghans, A. R., & Voss, B., (2000). Relativistic radioactive beams: A new access to nuclear-fission studies. *Nuclear Physics A, 665*(3, 4), 221–267.

113. Schunck, N., & Robledo, L. M., (2016). Microscopic theory of nuclear fission: A review. *Reports on Progress in Physics, 79*(11), 116301.

114. Serber, R., (1947). Nuclear reactions at high energies. *Physical Review, 72*(11), 1114.

115. Siegert, A. J., (1939). On the derivation of the dispersion formula for nuclear reactions. *Physical Review, 56*(8), 750.

116. Silberberg, R., & Tsao, C. H., (1973). Partial cross-sections in high-energy nuclear reactions, and astrophysical applications. I. Targets with z<= 28. *The Astrophysical Journal Supplement Series, 25*, 315–333.

117. Sitek, A., & Celler, A. M., (2015). Limitations of Poisson statistics in describing radioactive decay. *Physica Medica, 31*(8), 1105–1107.

118. Smith, A. M., (1952). LXXXIX. On the theory of beta-radioactivity. *The London, Edinburgh, and Dublin Philosophical Magazine and Journal of Science, 43*(344), 915–933.

119. Soddy, F., (1906). The present position of radioactivity. *Journal of the Röntgen Society, 2*(7), 45–65.

120. Specht, H. J., (1974). Nuclear fission. *Reviews of Modern Physics, 46*(4), 773.

121. Strauch, K., (1951). High energy photo-nuclear reactions. *Physical Review, 81*(6), 973.

122. Studier, M. H., & Hyde, E. K., (1948). A new radioactive series—The protactinium series. *Physical Review, 74*(5), 591.

123. Tamplin, A. R., (1972). *A Proposal for a Rational Policy to Control Radioactivity and Other Forms of Pollution* (p. 87). .

124. Thomas, A. W., (1984). Chiral symmetry and the bag model: A new starting point for nuclear physics. In: *Advances in Nuclear Physics* (pp. 1–137). Springer, Boston, MA.

125. Thompson, I. J., (1988). Coupled reaction channels calculations in nuclear physics. *Computer Physics Reports, 7*(4), 167–212.

126. Toschi, R., (1997). Nuclear fusion, an energy source. *Fusion Engineering and Design, 36*(1), 1–8.

127. Trautmann, W., Bianchin, S., Botvina, A. S., Le Fevre, A., Leifels, Y., Sfienti, C., & Aladin, Collaboration, (2010). The symmetry energy in nuclear reactions. *International Journal of Modern Physics E, 19*(08n09), 1653–1663.

128. Van, B. J., (1985). The possible influence of the discovery of radioactive decay on the concept of physical probability. *Archive for History of Exact Sciences, 31*(4), 369–385.

129. Vogt, E., (1962). Theory of low energy nuclear reactions. *Reviews of Modern Physics, 34*(4), 723.

130. Voltaggio, M., Palmisano, M., Raponi, A., & Voltaggio, S., (2001). Implantation of recoiling radionuclides of U and Th radioactive series applied to estimation of surficial erosion of $CaCO_3$ materials. *Applied Geochemistry, 16*(7, 8), 835–848.

131. Wahl, W., (1941). A new radioactive series. *Science, 93*(2401), 16, 17.

132. Weidenmüller, H. A., & Mitchell, G. E., (2009). Random matrices and chaos in nuclear physics: Nuclear structure. *Reviews of Modern Physics, 81*(2), 539.

133. Weinert, F., (2009). Radioactive decay law (Rutherford–soddy). In: *Compendium of Quantum Physics* (pp. 630–632). Springer, Berlin, Heidelberg.

134. Weisskopf, V. F., & Ewing, D. H., (1940). On the yield of nuclear reactions with heavy elements. *Physical Review, 57*(6), 472.

135. Weisskopf, V., (1937). Statistics and nuclear reactions. *Physical Review, 52*(4), 295.

136. Winter, J., (2000). Dust: A new challenge in nuclear fusion research?. *Physics of Plasmas, 7*(10), 3862–3866.

137. Woosley, S. E., Heger, A., Cumming, A., Hoffman, R. D., Pruet, J., Rauscher, T., & Wiescher, M., (2004). Models for type I x-ray bursts with improved nuclear physics. *The Astrophysical Journal Supplement Series, 151*(1), 75.

138. Zink, D., Fischer, A. H., & Nickerson, J. A., (2004). Nuclear structure in cancer cells. *Nature Reviews Cancer, 4*(9), 677–687.

CHAPTER 6

COSMOLOGY AND MODERN ASTROPHYSICS

CONTENTS

6.1. INTRODUCTION

The Big Bang, according to most scientists and astronomers, is the origin of our universe. We will show several of the quantitative data that support this notion in this section. We have delayed till now to teach cosmology since it is dependent on the theory of relativity and seems to have significant connections to basic particle physics. Our understanding of astrophysics is growing, particularly in the last 20 years. The field is part of the unexplainable data and contradictory ideas in this section. Cosmology and current astrophysics are inextricably linked, and most of what we provide in this section is about both (Shore, 2002; Carroll and Ostlie, 2006).

The extent to which cosmology, as well as astrophysics, are dependent on other domains of physics may surprise students when they first encounter these topics. Quantum physics and gravity have a starring role. To discover how key aspects were produced in the early phases of the cosmos, measurement methods of nuclear cross-sections are required to comprehend nucleosynthesis. Elementary particle physics and astrophysics have proceeded to combine to the point that the disciplines now intersect significantly, as we shall show in the next part on the Big Bang. Cosmologists and elementary particle physicists both have a large passion for the basic forces. The test data of the future can invalidate today's deeply held ideas. Earth- and space-based observatories are answering our most baffling questions, and new acceleration tests may tell us a lot much more about the planet's beginning (Sergii et al., 2018; Kramar and Ilchenko, 2021).

6.2. EVIDENCE OF THE BIG BANG

In the last half-century or more, scientists suggested several ideas about the beginning of the universe. In the 1950s and 60s, there were two competing cosmological hypotheses, but none was supported by adequate evidence. Both hypotheses explained the observed rate of expansion, which was convincingly confirmed in 1929 by Edwin Hubble's redshift measurements. One idea, called the steady-state theory, maintained that material is continually generated; while the cosmos grows, its density remains steady (Robertson, 1933; Kardashev, 1997). The alternative hypothesis, the Big Bang model, posits that the world was born in an incredibly dense and hot primordial fireball. In the next part, increasing data in favor of the Big Bang models would be provided. In discussing the Big Bang model, we give three significant sources of evidence that have persuaded the majority of the scientific community to regard this as the most plausible account of the beginning of the universe. These are the lines of evidence:

- From 1929 to 1952, utilizing the huge telescopes on Mount Wilson and Mount Palomar, Edwin Hubble observed that such galaxies of the cosmos are migrating apart at a rapid rate. The cosmos seems to be growing from certain primordial events.
- The 1964 finding by Arno Penzias and Robert Wilson of Bell Laboratory that a microwave amount of radiation penetrates the cosmos. This background radiation is due to the Big Bang.
- There is a strong correlation between both the predictions of primordial nucleosynthesis and the observed quantity of components in the cosmos. This pertains to the light components created in the early phases of the Big Bang (Flin and Duerbeck, 1989; Larson, 2005).

Although many of the expectations and outcomes of the Big Bang model are still contested, its core concepts are universally acknowledged. Because it was initially presented, it has undergone several modifications, as we will see in the next sections when we address a few of the obstacles. The Hubble expansion of the cosmos, the cosmic ray's radiation, as well as the relative amounts of photons, are examined in depth here.

6.3. HUBBLE'S MEASUREMENTS

Edwin Hubble's amazing observations and comprehension of our cosmos span 40 years. In the middle of the 1920s, he demonstrated that there were galaxies apart from our own Although his groundbreaking study contradicted other published evidence at the time, Hubble was capable of persuading others to embrace his theories by methodically and painstakingly making a compelling case (Deligny and Suomijaervi, 2015; He et al., 2019).

The change to shorter wavelengths (redshift) of particular spectral lines generated by distant objects is used to calculate the recessional speed of celestial objects. Several objects' recessional velocities were determined by Hubble. Several astronomers had observed that some nebulae looked to be retreating at large elastic velocities from us, most notably V. M. Slipher of Lowell Observatory in 1912, although it was Hubble who placed this experimental reality on a solid foundation. The connection called Hubble's law relates the speed v with which galaxies recede from us to the distance R from Earth (Figure 6.1):

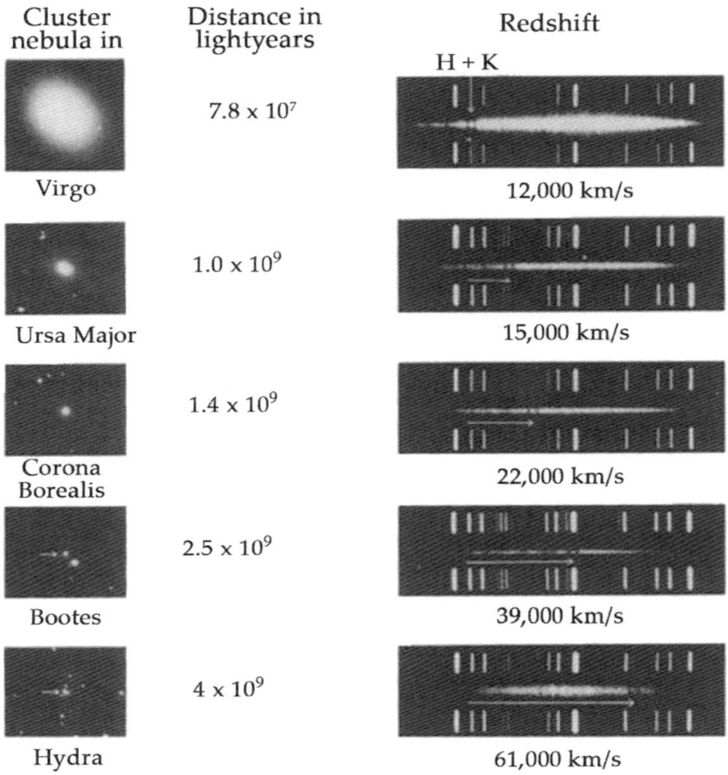

Figure 6.1. Redshift data for various galaxies are shown with their distance from Earth in light-years. The spectrum of each galaxy is the wide, hazy band placed in the middle between the laboratory spectra contrast. The calcium absorption lines K (393 nm) and H (397 nm) are redshifted by the arrowhead as well as the shift to the right for greater velocities. These preliminary findings are compelling and demonstrate that the cosmos is expanding.

Source: https://astronomy.com/magazine/ask-astro/2018/05/rotational-speed-of-a-galaxy.

v = HR (1)

The Hubble parameter H is connected to a multiplier that is equal to the difference between galaxies by a factor of a.

$$H = \frac{1}{a}\frac{da}{dt} \qquad (2)$$

Hubble's parameter does not stay unchanged as the cosmos expands, rather it gradually diminishes over time. It is frequently abbreviated as H_0 and referred to as the Hubble constant. In 1927, George Lemaître used general relativity to develop Hubble's law, although his discovery was not well recognized since it appeared in a Belgian publication It is entitled to understand the distance R to objects whereby the redshift has indeed been calculated to evaluate if Hubble's law is correct (Beskin, 2007; Chopra et al., 2017). Hubble created complex approaches for determining distances based on the brightness of stars and galaxies with some extra assumptions, R. Hubble was capable of achieving this with several accuracies for stars out to distances of 10 million light-years as well galaxies out to distances of 500 million light-years. Hubble and his brilliant colleague Milton Humason studied hundreds of star sources over several years to establish their redshifts (Figure 6.2) (Lemoine and Sigl, 2001).

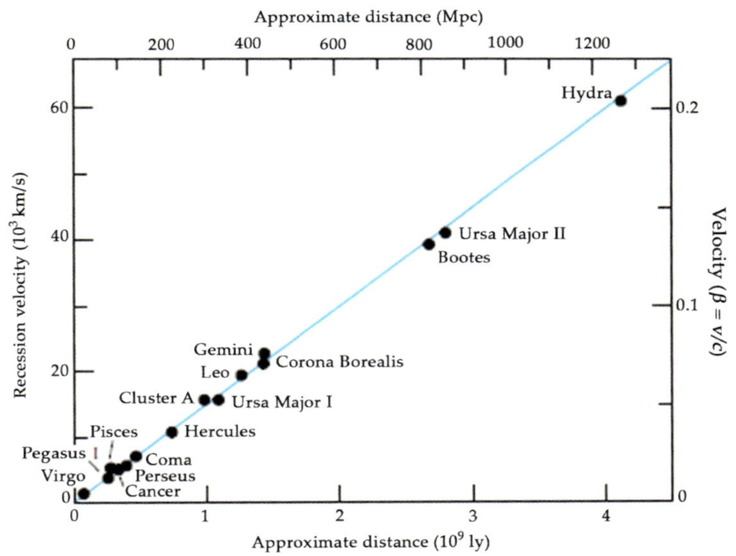

Figure 6.2. The recession speed as a function of distance for 15 clusters of galaxies.

Source: https://www.researchgate.net/figure/Plot-of-recession-velocity-and-istance-in-Mpc-Courtesy-official-web-page-of-University_fig4_318259770.

He demonstrated that 49 galaxies (dubbed extragalactic nebulae) match the velocity-distance relationship in 1929. Hubble's findings needed meticulous brightness and redshift measurements. Hubble and Humason had cataloged the redshifts of 100 more galaxies by 1935, and the information unmistakably demonstrated that the galaxies furthest distant from us were traveling at the fastest rates.

Hubble demonstrated that Eqn. (1) is correct. Even though the distance measurements have been revised throughout the years as additional data has been acquired, the linearity of v and R holds today (Figure 6.2). Hubble's constant H_0 is now estimated to be about 22 km/s per million light-years (Kurt, 1997; Smith, 2017).

To see the extension, Earth does not need to be in the middle of the galaxy. Figure 6.3 depicts a balloon with dots. As the balloon inflates, all of the dots get more distant from one another. The top of the balloon is two-dimensional; currants in bread dough are a three-dimensional illustration often used. The currants split as the bread swells and stretch in three dimensions during baking. The raisins spread out farther, with those on the edges going quicker. Even as the cosmos expand and galaxies split, anything similar occurs (Izotov and Thuan, 2010; Tanndal and Waldenström, 2018). The idea of utilizing Hubble's constant to establish the origin of the universe would be discussed in the following sections.

Figure 6.3. A picture of how galaxies are retreating from one another. Each dot gets farther away from the other dots as the balloon inflates. The cosmos seems to stay homogenous as it expands.

Source: https://opentextbc.ca/universityphysicsv3openstax/chapter/the-big-bang/.

6.4. COSMIC MICROWAVE BACKGROUND RADIATION

In 1964, Arno Penzias and Robert Wilson investigated radio transmissions in the microwave wavelength band. Their specifically built low-noise antenna continued to grab an obnoxious, persistent signal. Penzias and Wilson were at a struggled to explain their conclusion after numerous efforts to eradicate the unwanted noise failed. In a chat with some other colleagues, they learned that P. James E. Peebles and the team at Princeton had projected a Big Bang-era radiation environment. Robert Dicke, an additional Princeton physicist, was preparing a test to detect it. Penzias and Wilson noticed this residual background radiation (Tacchella et al., 2015; DeFond, 2019).

Due to the fast-expanding and freezing of the cosmos, protons, and electrons eventually formed atoms that photon radiation can no longer ionize. At a temperature of 3,000 K, matter dissociated from radiation. This 3,000 K blackbody radiation has Doppler-shifted to a maximum of around 3 K. The redshift is around a factor of 1,000, depending on the exceedingly high velocity of that region of the cosmos relative to the planet. Figure 6.4 depicts the blackbody radiation spectrum. Subsequent satellite observations have revealed that the total blackbody spectrum is almost isotropic, although not entirely so. We shall revisit this topic later (Feduccia, 2003; Gurzadyan and Penrose, 2010).

The measurement of the cosmic background radiation by Penzias and Wilson (Nobel Prize in Physics, 1978) was a win for the Big Bang hypothesis, and the finding was extremely tough to describe with the steady-state model. After the first discovery, it became known publicly that George Gamow, Ralph Alpher, and Robert Herman had calculated in the late 1940s and early 1950s that such Big Bang leftover radiation could occur in the range of 5–7 K (Ringwald, 2005; Tilvi et al., 2020).

6.5. NUCLEOSYNTHESIS

The cosmos had calmed sufficiently a few moments just after Big Bang that neutrons and protons may undertake thermonuclear fusion as well as generate particles. The cosmic microwave background radiation gives us a look back in time to around 400,000 years after the Big Bang, yet element production only took a few minutes. The nuclear fusion of elements thus serves as a thorough examination of the Big Bang scenario (Chen et al., 2017; Freedman, 2021). Using the current relative existence of the elements, physicists may test the conditions of the cosmos that have occurred as early as a fraction of a second after the Big Bang.

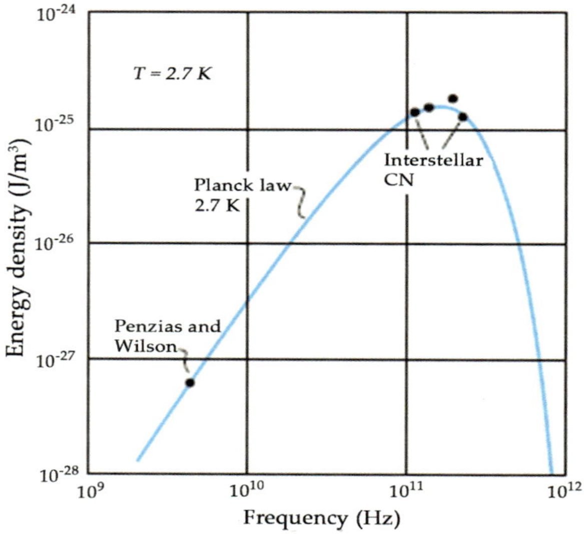

Figure 6.4. Calculated blackbody radiation distribution is shown as a function of frequency; the datum point that Penzias and Wilson measured is noted. The first measurement of the cosmic microwave background was probably made by Andrew McKellar in 1940.

Source: https://en.wikipedia.org/wiki/Cosmic_microwave_background.

The light particles 2H, 3He, 4He, and 7Li are generated in reactions such as the proton-proton cycle, so expectations from the Big Bang model may be contrasted to current data. Although heavier atoms are created in stars, hydrogen and helium make up the overwhelming bulk of the known mass in the cosmos (Wu et al., 2017; Mukherjee et al., 2021).

Some other elements are found in trace amounts. In the 1960s and 1970s, results were calculated utilizing our understanding of nuclear and particle physics to contrast with research observations of elemental quantities. Figure 6.5 shows that the measurements and Big Bang predictions for a density of roughly 3×10^{-31} g/cm^3 were very similar. This is referred to as the universe's normal material density, which is mostly owing to baryons. In the next part, we'll examine how current data suggests that most of the substances in the world has yet to be discovered.

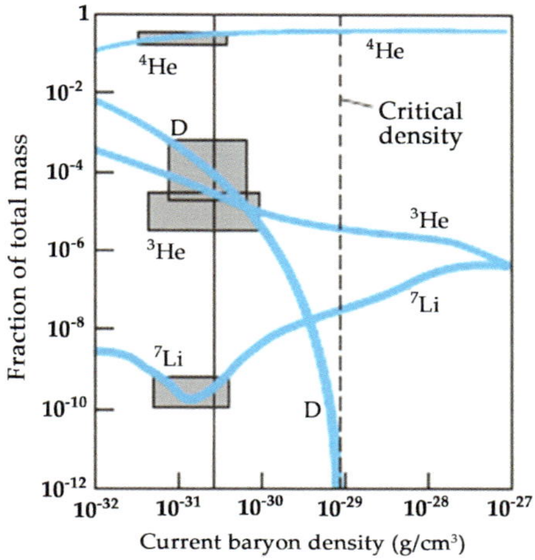

Figure 6.5. The fractional mass abundances of several light elements are displayed against possible current baryon mass densities. The boxes represent experimental observations, and the solid curves are calculations of the standard model of the Big Bang.

Source: https://www.mdpi.com/2073-8994/14/3/518/htm.

One version of the Big Bang theory predicts an inflationary cosmos, with an exact solution of roughly 9×10^{-30} g/cm^3. 4He makes up around 25% of the universe's known (or conventional) mass, with the rest being virtually entirely of unbound protons. During the early phases of the cosmos, the proportion of neutrons to protons and the quantity of material were important in the formation of 4He.

6.6. THE BIG BANG

The Big Bang framework is characterized by two theoretical pillars:

- The generalized theory of relativity; and
- The cosmological principle;

The cosmological principle hypothesizes that the cosmos appears about identical anywhere and in all directions. The cosmos would be both isotropic and homogenous. In 1922, the Russian mathematician Alexander Friedmann utilized such two foundations to establish the development of the cosmos as described by his Friedmann equations. From 1927, when Georges Lemaître separately came to conclusions comparable to Friedmann's, the cosmos has been getting larger, as suggested by the answers to his equations. Then, in 1935, A. G. Walker and H. P. Robertson separately demonstrated that the Robertson-Walker metric is the smallest spacetime model compatible with an isotropic, homogeneous world. The temporal development of spacetime is defined as a simple variable that varies with time per the Friedmann equation [read Eqn. (3) bottom] (Hubble et al., 2009; Baxter and Sherwin, 2021). These words are known by a variety of names, such as Friedmann, Lemaître, Robertson, and Walker (for instance, FRW or FLRW universes). The concept is congruent with the growing balloon dotted with galaxies seen in Figure 6.3.

A few of the Friedmann cosmological models may be written down,

$$(\frac{1}{a}\frac{da}{dt})^2 = \frac{8Gpm}{3} - \frac{kc2}{a2} + \frac{c2}{3} \qquad (3)$$

wherein; G is the universal gravitational standard, a is the shape factor added in Eqn. (2), pm is the average weight concentration of the galaxy, k is the universe's curvature factor, and is the cosmological constant given to the initial Friedmann equation. Einstein proposed the factor (lambda) to establish a static world and a homogeneous cosmos since astronomers convinced him that the world was not in movement. The cosmological constant term accounts for the energy of perfect vacuum in order to have the homogeneous and isotropic universe. (Wu et al., 2020; Philcox et al., 2021). When Hubble gave testimony for the extension of the cosmos, Einstein suggested the cosmological constant as his worst error. Several years back, cosmologists often set = 0, but now it is accepted that the cosmological constant has a particular benefit and is related to the power output of the cosmos.

We modify Eqn. (3) by utilizing the Hubble parameter H from Eqn. (2).

$$(H)^2 = \frac{8Gpm}{3} - \frac{kc2}{a2} + \frac{c2}{3} \qquad (4)$$

This is commonly referred to as the Friedmann equation and is crucial to cosmology. If we can discover the missing parameters in the equation, we may be able to establish both the weight and health of the universe. Either part of the equation is divided by H_2 to get:

$$1 = \frac{8Gpm}{3} - \frac{kc2}{a2} + \frac{c2}{3} \qquad (5)$$

The critical concentration of the cosmos is the density where the universe cannot grow or shrink. In other words, it generates a flatter universe, which appears to be the situation now. By fixing the radius of curvature parameter k = 0 and assuming = 0 in equation, we can get the critical density (Eqn. (5)).

Then, we have:

$$p_c = \frac{3H2}{8G} \quad (6)$$

The density factor is calculated as the proportion of the real universe density to the exact solution.

$$\Omega = \frac{p}{pc} = \frac{8Gp}{3H2} \quad (7)$$

The current density of the cosmos is far below the critical density. In addition to baryons, there are undetected additions to density from dark energy and dark matter. To explain for actual observations, it is expected that the density factor is near 1 (Farooq et al., 2013; Momcheva et al., 2016). The three variables in Eqn. (5) are unitless, as is the density factor o. We shall demonstrate that it is easy to rephrase Friedmann's Eqn. (5) as 1 = = m + k +. The first component is the mass density m, which contains both baryons and dark matter. The second component is related to the curvature k, while the third term indicates the vacuum energy density Ω_Λ.

$$\Omega_m = \frac{8Gpm}{3H2} \text{ Mass density} \quad (8a)$$

$$\Omega_m = \frac{8Gpm}{3H2} \text{ Mass density} \quad (8b)$$

$$\Omega_\Lambda = \frac{c2}{3H2} \text{ The vacuum energy density} \quad (8c)$$

In this amazing condition, m, and dictate the development of the cosmos. Ever since the 1960s, the Big Bang has been commonly recognized as the beginning of the universe. Our understanding of nuclear and basic particle physics and the theory of relativity enables us to comprehend the evolution of the cosmos from before the Big Bang. In this part, we define t = 0 as the origin of the world, or the Big Bang, and outline the significant events that have transpired in the ensuing 13.7 billion years. Figure 6.6 depicts the temperature of the cosmos as time-dependent. $t = 0 \rightarrow 10^{-43}$ seconds We have no hypotheses that can explain what occurred during this period, termed the Planck epoch since the existing physical rules are not enforced (Morse et al., 2001; Addison et al., 2018). Most probably, the cosmos began with an unlimited mass density and zero spacetime curvature. This circumstance is referred to as a cosmological singularity. In 10^{-43} s, the extent of the observable cosmos was likely less than 10–52 meters. Almost certainly, the temperature of the stuff exceeded 1,030 Kelvin. If the four forces of strong, electromagnetic, weak, and gravity were merged into one force, our current theories of gravitation and quantum physics will no longer be applicable. Only a theory of gravitation that incorporates quantum physics, or quantum gravity, would have any chance of understanding what occurred. There are currently no credible options for this cause, although it has been termed the "theory of everything." $t = 10^{-43}$ s → 10^{-35} s As bizarre as it may sound, we feel we get some grasp of what occurred throughout this age. By 10^{-35} s, the cosmos had grown to a diameter of around 10^{-35} m and the temperature had decreased to maybe 1,028 K. Gravity was identified as the first distinct force. This age is referred to as the Grand unification Era, and the grand unification theories (GUTs, involving electroweak and strong forces) explain it, although none have been effective as of yet (Bolejko, 2018; Birrer and Treu, 2020). $t = 10^{-35}$ s × 10^{-13} s Even before the Electroweak age begins; the strong force had split off and formed a distinct force. The basic elements (quarks and leptons) and their antimatter particles have been created. The cosmos is a heated quark-electron soup. The cosmos proceeded to cool, maybe reaching 1,016 K in 10^{-13} s. Currently, no accelerator has sufficient particle energy to investigate what occurred at such high temperatures.

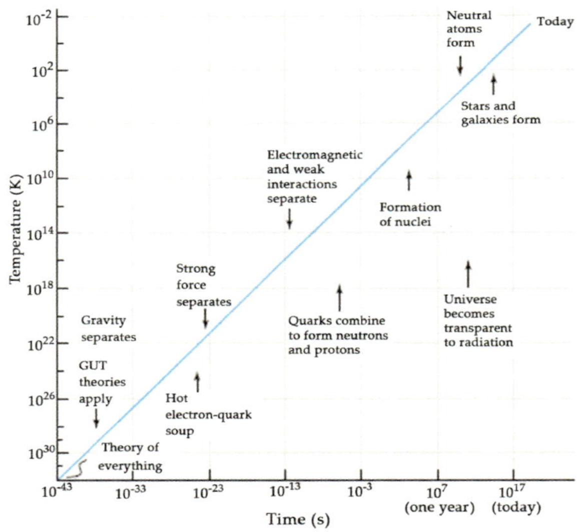

Figure 6.6. The temperature of the universe is displayed as a function of time since the Big Bang.

Source: https://www.researchgate.net/figure/The-temperature-evolution-of-the-Universe-as-a-function-of-time-in-the-early-stage-of-the_fig1_26328935.

Currently, only a few cosmic rays possess similar energy. Throughout this time, the dimension of the cosmos changed by multiple magnitudes and reached a size of 10 cm. $t = 10^{-13}$ s $\rightarrow 10^{-3}$ s Throughout this Quark period, quarks fused to produce neutron and proton Just the two lowest particles, down, and up, were efficient in the thermodynamics of the cosmos, because the more substantial quarks had low speeds at this epoch that they might not transmit any energy. Since protons are somewhat less heavy than neutrons, there have been more protons than neutrons. At energies exceeding 100 GeV (equivalent to temperatures of around 1,015 K), electromagnetic, and light attractions were united into the electroweak attraction (Jackson, 2015; Davis et al., 2019). Nevertheless, beneath 100 GeV, the W' and Z bosons act as heavy particles, while the photon is extremely small, hence the electromagnetic and light attractions have lost their equilibrium and function independently. We claim "symmetry is shattered" below 100 GeV. After this time, the temperature had plummeted to around 1,011 K. The four factors of now are now distinctive. By the conclusion of this epoch, the cosmos was comprised mostly of photons, electrons, neutrinos, protons, and neutrons, and also different antiparticles (for example, positrons, and antiprotons). The observable universe was around 1,000 meters. By the conclusion of this time, the cosmos had cooled sufficiently (to 109 K) that such photons had insufficient energy to dissolve newly produced deuterons (Chen et al., 2015; Basilakos et al., 2018). This marked the start of nucleosynthesis or the production of nuclei. Helium as well as other light atomic nuclei were created during nucleosynthesis (for instance, the proton-proton cycle). As a result of neutron decay, there were significantly more protons as compared to neutrons. After this age, the cosmos had cooled to as little as 104 K. The cosmos was predominantly composed of photons, protons, helium nuclei, and electrons. Atoms could not form since the powerful electromagnetic radiation ionized these practically immediately after formation. Photons easily reacted with electric charges via electromagnetic force and were accepted, produced, and dispersed by matter (Wilson, 1979; Gawiser and Silk, 2000).

From t = 300,000 years to the present, the cosmos had eventually cooled sufficiently for electromagnetic radiation (photons) to become disconnected from matter. Till around 300,000–700,000 years ago, the majority of the energy in the cosmos was in the kind of photons, that were continuously received and released by ions. At around 3,000 Kelvin, the temperature was sufficiently low protons may unite with electrons to generate neutral hydrogen atoms. At this time, the dispersion of light from neutral hydrogen (as compared to unbound protons and electrons) decreased considerably, and electromagnetic radiation was able to wander across the cosmos. Since photons may now freely travel across the cosmos, a 3,000 K blackbody radiation will endure forever (Schlegel et al., 1998; de Bernardis et al., 2000). The great bulk of this 3,000 K radiation is redshifted relative to photons in the cosmos today are a result of the 3 K background. This is the cosmic microwave background radiation explained in the past segment, and it could be caught by our televisions. It is a component of the snowy image produced by an antenna while no powerful channel is obtained. The matter started to cluster with others to make proteins, gas clouds, stars, and ultimately galaxies after the formation of atoms. The rest is now history! From this point on, the universe grew into its current shape. The cosmos is now driven by matter, with much more energy to produce matter as compared to radiation (Wright et al., 1992; Partridge and Partridge, 1995).

6.7. STELLAR EVOLUTION

Matter in the type of electrons, protons, and 4He traveled across the cosmos 4 billion years now since the Big Bang, somewhat like air molecules in a vast room. Gravitational forces ultimately decided to muster a few of the stuff combined with large gaseous clouds, that provided the core for stars, as the temperature decreased over time. The KE of the protons increased as they were pulled together by their gravity contact. As the core temperature of such young stars increased, this activity persisted. A gaseous cloud's core had a greater concentration and heat than the exterior. The cloud proceeded to compress till the temperature hit around 107 K, signaling the start of the nuclear fusion cycle. A star's feature is nuclear fusion. It might have required a million years more than for the clouds to compress enough to trigger fusion, allowing the star to form, however, some new findings suggest it could occur possible in as short as 200,000 years. Figure 6.7 depicts the conceptual development of a star (Tsallis et al., 1995; Srianand et al., 2000).

We explored nuclear fusion as a type of power source in stars. Radiation is produced when the proton-proton chain produces heat. This fusion system gradually produces 4He, which gathers in the star's core. If the temperature in the star is great sufficiently, other mechanisms generate 12C and heavier masses; this happens later in a star's existence. Of course, we are the most knowledgeable of our star, the sun. The color of emitted light may be used to determine the surface heat of stars; however, the internal temperature is harder to determine. The sun's surface temperature is estimated to have been about 5,800 K, with the body temperature reaching 14,106 K (Wagoner, 1973; Boesgaard and Steigman, 1985). A star on the scale of our sun might survive for 1,010 years, but a bigger star's nuclear fuel would deplete far more quickly. The light that we now get was most likely created over 105 decades previously in the sun's inner and then scattered numerous times before being radiated from the surface (White et al., 1994; Cyburt et al., 2016).

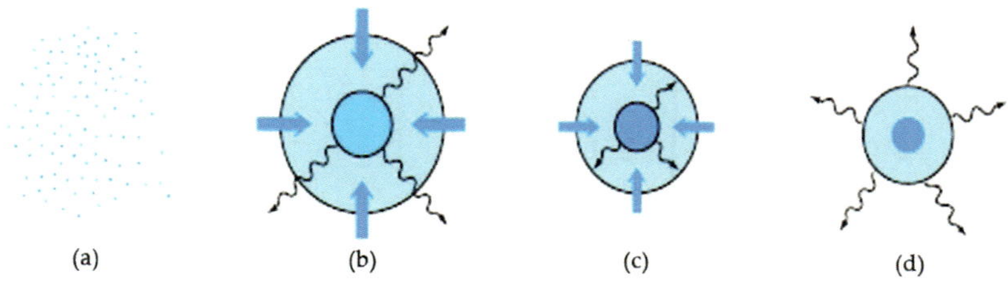

Figure 6.7. (a) Interstellar gas as well as clouds compact due to gravitational pull, forming stars. (b) A core is formed when the substance compresses, heating up and radiating energy. (c) Ultimately, the outside area gets so thick that the heated inner core's rays cannot leave. The fall diminishes, however the matter proceeds to heat, resulting in the formation of a protostar with a high concentration and temperature. (d) The shrinkage of a star of around 1 solar mass (our sun's size) finally warms up sufficient to support nuclear fusion. The gravity contraction is balanced by the blackbody radiation generated by nuclear fusion, and the star remains stable as a main-sequence star. For hundreds of millions of years, the star would fire, turning hydrogen into helium.

Source: https://www.britannica.com/science/star-astronomy/Star-formation-and-evolution.

6.8. THE ULTIMATE FATE OF STARS

The ultimate phases of star life occur whenever the hydrogen fuel in the center runs out. At this stage, gravitational pull remains, while concentration and temporary rise. The temperature rises to the point when helium nuclei start to combine. Including well nuclear fusion procedures, massive elements are production represents (Hoyle and Fowler, 1960; Schramm and Turner, 1998). For more star formation, the fusion mechanism proceeds till nuclei close to the iron area are generated, wherein elements have the maximum binding energy (BE) for each nucleon. The nuclear fusion cycle could not proceed, and all reactions cease. The eventual destiny of the star relies on its mass. For stars considerably more powerful as compared to the sun, the gravitational influence of the mass remains till the star's possible to reach an incredible level. Examine this procedure in further depth (Hughes et al., 1999; Kawasaki et al., 2008). Let there be N nucleons, every with mass m, in the star. The gravitational self-potential power of a spherical with mass Nm and radius R,

$$\text{P.E.} = U_{grav} = -\frac{3}{5}\frac{G(Nm)2}{R} \tag{9}$$

wherein; U is being used for potential energy (PE) instead of V to clarify things with volume V. We calculate the gravity pressure by calculating the force per unit of area.

$$P_{grav} = \frac{3}{5} = -\frac{1}{4R2}\frac{d\,\text{Ugrav}}{dR} = \frac{3G}{5}\frac{(Nm)2}{4R4} \tag{10}$$

We express gravitational pressure in terms of volume using V = (4)/(3) R3 to get:

$$P_{grav} = 0.32G\,\frac{(Nm)2}{V4/3} \tag{11}$$

The outer pressure of electrons prevents the entire disintegration of matter. This arises since this Pauli exclusion principle prevents two electrons from holding the identical state. Nevertheless, for a suitably big star, gravity would ultimately compel the electrons to collide with the protons:

$$e^- + p \rightarrow n + v_e \tag{12}$$

Neutron stars are most often comprised of neutrons. Nevertheless, since neutrons also adhere to the exclusion rule, neutrons will likewise produce an outward pressure compared with that of electrons (Maeder and Lequeux, 1982; Walker et al., 1991). Utilizing methods established, it is proved that the density of an electron gas equals:

$$P_e = \frac{2f2}{3}\frac{\hbar 2}{2me}\left(\frac{Ne}{V}\right)^{5/3}$$

wherein; f is the proportion of EF that correlates to total power. The energy content of a Fermi gas is around (3)/5 EF, hence we may deduce that f 3/5.

We can use the same relation for neutrons by replacing N_e by N and m_e by m. The neutron pressure becomes. Neutron pressure reaches:

$$P_n = \frac{3.9\hbar 2(N/V)5/3}{2m} \qquad (13)$$

In balance, the neutrons' outward stress owing to the exclusion concept will equalize the gravitational stress. We equalize Eqns. (11) and (13):

$$0.3G\,\frac{(Nm)}{V3/4} = \frac{3.9\hbar 2(N/V)5/3}{2m} \qquad (14)$$

We calculate this equation to get the quadratic formula of the volume:

$$V^{1/3} = \frac{6.5\hbar 2}{N1/3m3G} \qquad (15)$$

Notice that the diameter of a neutron star decreases as its mass increases. In the next example, we utilize this technique to determine the diameter of neutron stars (Meyer and Meyer-Hofmeister, 1999; Hujeirat, 2018).

6.9. ASTRONOMICAL OBJECTS

Galaxies are clusters of stars held together by the gravitational pull of the stars. The Milky Way, our particular galaxy, is thought to be made up of two trillion stars. There are at minimum one trillion galaxies in the universe (1011). There was a lot of debate in the 20th century over what the Milky Way was as a whole of the cosmos (Islam, 1977; Hogan, 1994). Astronomers introduced the word nebula to objects with a hazy, unsettled look. Hubble discovered in 1924 that Andromeda Nebula is a distinct galaxy that is 106 light-years distant from us. The Andromeda Galaxy is a fitting name for it. While several, tiny galaxies are nearer to our Milky Way, the Andromeda Galaxy is the closest and may be seen with the naked eye in a clear sky. Astronomers are discovering more distant galaxies (Smith and Owocki, 2006; Kirsebom et al., 2016). Numerous galaxies with redshifts of 10 or greater have been discovered. We're witnessing light from such galaxies that have been created while they were just 500 million years old when they were still in their childhood. These galaxies are rapidly traveling away from us, and their light has gone almost 13,000,000,000 years to hit us. Astrophysicists and cosmologists are fascinated by the development of galaxies within the early cosmos (Bechhoefer and Chabrier, 1993; Bruenn and Mezzacappa, 1998).

6.9.1. Active Galactic Nuclei (AGN) and Quasars

In the 1960s, astronomers detected entities with very powerful radio emissions and incomprehensible visual spectra. Due to their huge redshifts, Maarten Schmidt demonstrated in 1963 that such entities must be at minimum 3 billion light-years distant from Earth. He was capable of understanding the hydrogen's shifted spectral lines. These phenomena have been named quasars, which stands for quasi-stars or quasi-stellar objects, and they may outshine galaxies with tens of trillions of stars (Appenzeller, 1967; Wenger et al., 2000).

Nowadays hundreds of quasars have been identified, it is known as they produce light across the entire electromagnetic spectrum, from gamma rays to radio waves, including significant optical and ultraviolet elements. Approximately 10% or less of quasars release intense radio emissions. Scientists hypothesized that only a supermassive black hole at the heart of a galaxy can generate this tremendous energy (Dayton et al., 2002; Yurchenko et al., 2014). Quasars are the most distant things in the universe but were consequently one of the first objects to originate in the cosmos, approximately 12 billion years back. As a result of the immense range between the galaxy and the observer, the host galaxy is obscured by the considerably strong light of the quasar. The reality that the luminosity of quasars may change in just some hours or days implies that the diameter is just several light hours and light days wide, which is not much bigger than our universe (Zhang and Zhao, 2004; Berkhout and Beijersbergen, 2008).

Whenever infant galaxies were developing ages ago, the central stars were tightly concentrated. These stars' crashes and merging led to the birth of supermassive black holes. Such black holes needed enormous quantities of mass fuel, that was supplied by gas from the interstellar medium, a neighboring galaxy, or a too near star (Michelson, 1920; Lang et al., 2016). Seen at immense ranges, quasars were likely generated eons ago. We do not observe quasars near our planet; hence the current circumstances are not suitable for the development of quasars. Consequently, quasars must develop into things that are prevalent today, very probably regular, inactive galaxies. If this is true, then certain regular galaxies may contain huge black holes which are no more quasars since they have run out of fuel. We need not realize that the majority of regular galaxies went through the quasar stage (Camenzind, 2007; Axelrod et al., 2010).

In the 1970s, unraveling the enigma of quasars was deemed so crucial that this was a primary reason for the Hubble Space Telescope (HST).

Spectacular photos from the HST have confirmed that quasars live inside galaxies. Nonetheless, the photos have shown more unexpected findings. There is a wide range of galaxies; a few are regular, while others are combining and merging with neighboring galaxies, drawing their material away. We currently think that the supermassive black holes at the core of quasars switch on and off as they absorb mass distributions from their galaxies or as galaxies collide. Although quasars remain a mystery, they give us a glimpse into the circumstances that existed early in the universe's existence (Morrison, 1958; Ashai et al., 2022). An astounding variety of strange extragalactic objects are classified as active galactic nuclei (AGN). Which involve blazars, Seyfert galaxies, and very bright quasars. We hypothesize that their center comprises a supermassive black hole encircled by a mass-supplying accretion disk. As matter spirals into touch with a black hole, enormous quantities of electromagnetic radiation and plasma jets are ejected into space, often in directions perpendicular to the accretion disk. Figure 6.8 depicts an artist's rendition of this topic. By seeing the jets in a straight line, we observe extraordinary brightness (Sulentic and Tifft, 1979; Pino et al., 2008).

Blazars are elevated galactic nuclei with relativity jets that emit energy straight at Earth. They were founded in 1991 to use the Compton Gamma Ray Observatory. They are accompanied by powerful radio transmissions.

Figure 6.8. An artist's rendition of a galactic nucleus is seen here. It is thought that there is a supermassive black hole in the heart of the core, which is exceptionally brilliant. Gas from interstellar space of the galaxy, from stars that wander too near, and from neighboring galaxies drop into the black hole.

Source: https://www.popsci.com/black-hole-early-universe-stars/.

6.9.2. Gamma-Ray Astrophysics

GRBs (gamma-ray bursts) are brief pulses of electromagnetic radiation that occur every other day at irregular intervals and from various directions. The Vela satellites were initiated in the late 1960s to check the Soviet Union's conformity with the Nuclear Test Ban Treaty by detecting GRBs from nuclear detonations (Baxter, 1989; Carrasco-Davis et al., 2019). Vela did catch gamma bursts; however, they came from orbit rather than Earth, which was a startling revelation. Following the release of information of this data in 1973, scientists went out to figure out what caused these gamma-ray blasts, that might last anywhere from a few msec to many minutes. Because gamma rays are primarily soaked up by the Earth's atmosphere, space-based observatories have provided the majority of the information. Astronomers were unaware of where GRBs came from as late as the early 1990s. Did they originate in our universe, our Milky Way Galaxy, or somewhere else in the universe? GRBs occur from supernovae in galaxy clusters, according to a mixture of satellite and earth measurements and scientific estimates (Weigelt, 1977; Ochsenbein et al., 2000).

The "afterglow" of lower energy photons, such as x-rays, light, and radio waves, that lasts for weeks and even months is an intriguing aspect of GRBs. The radiation from a GRB travels in the path of the explosion, which could only be observed if the blast is aiming toward Earth. Observing the afterglow was critical to understanding the GRB phenomenon. The Italian-Dutch BeppoSAT spacecraft, which had both x-ray and gamma-ray monitoring capacity, was capable to identify and study the afterglows throughout its 1996–2002 study years, although the Compton Gamma Ray Observatory wasn't even prepared to detect them. Astronomers utilized BeppoSAT data to tell other astronomers where to point the optical telescopes in an attempt to obtain spectral data (Webster, 1991; Bonnarel et al., 2000). Astronomers were capable of determining redshifts of optical spectra in 1997 utilizing the powerful Keck and Hubble observatories, proving the bursts originated from beyond. The directional readings also indicated that the GRBs emanated from all over the place. GRBs have been recorded with luminosities estimated to be a magnitude 1,017 larger than our sun. The majority of explosions are currently thought to be radiation from a supernova's focused band (or jet) of radiation. A huge star's core explodes in a supernova explosion. A fast-moving blast wave or fireball clashes with stellar substances remaining within the star, resulting in gamma rays that explode out of the star immediately before the blast wave. The afterglow is caused by the blast wave colliding with dust and gas surrounding the star (Zhao and Zhang, 2008; Cabezas et al., 2022).

6.9.3. Novae and Supernovae

Novae and supernovae are also fascinating astronomical events. Novae are just stars that abruptly shine, are observable (hence the term nova, which means "new"), and afterward fade over time. It is said there are two kinds of supernovae, with Type I lacking hydrogen spectral lines and Type II possessing them. This classification method isn't any more practical, and Type I has been subdivided into Ia, Ib, and Ic based on other requirements (Figure 6.9).

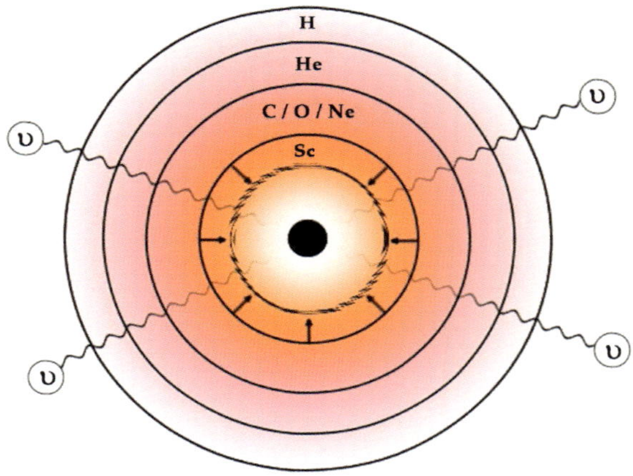

Figure 6.9. Supernova explosion.

Types Ib, Ic, and II share more characteristics than Type Ia. Type Ia supernovae are the smartest and are believed to happen in binary mixtures in which a white dwarf obtains matter from its partner till it could no longer support on its own and crumbles (Bates and Cady, 1980; Starck et al., 1995). As the main strong contributor elements, supernovae are essential. Lighter elements, like carbon and oxygen, are created by nucleosynthesis inside of stars and afterward ejected in the blast, where they could be assimilated through other stellar formations. However, atoms beyond iron could not be generated by nuclear fusion within stars. Extreme explosions in supernovae generate the heat and pressure necessary for the fusion of medium-mass elements into substances, such as uranium. Supernovae are also essential as light sources for measuring cosmic distances (Gustafsson et al., 2008).

Many supernovae might be observed by the Chinese in the last two millennia. There is extensive proof that now the outburst of the Crab supernova (Type I) happened in 1054. The Crab exploded less than 6,000 light-years from Earth, and that must be a spectacular sight. It was as brilliant as the planet Venus and visible throughout the day! The phenomenon was noticed and recorded by the Chinese, Japanese, and probably American Indians. The Crab Nebula, seen in Figure 6.9, has become a gorgeous cloud of blazing gas that may be viewed via a telescope. Tycho Brahe and Kepler spotted other supernovae in 1,572 and 1,604, respectively. Following 1,604 there does not appear to have been a supernova observable to the human eye till 1987. The unbelievable discovery of a supernova in 1987 substantially attracted great attention to supernovae (termed SN 1987A; see Figure 6.10).

An explosion of Supernova Astrophysicists now think they know how SN 1987A may have emitted more energy in 10 seconds thanas compared to the 100 of our suns could do that in 1,000 million years! A Type II supernova may be described in general terms. The specifics are fascinating, and so with the presence of a carefully monitored and quantified event, much was learned after SN 1987A. Gravity constricted the star to some other phase when hydrogen was depleted by thermonuclear fusion and the helium concentration grew. The helium nuclei started to fuse, making the star's central core denser and hotter. Helium, which had generated carbon and oxygen, eventually ran out. When the core temperature hit 8,108 K, carbon started to fuse, producing neon, magnesium, and silicon nuclei. Most of the nuclei eventually merged to create nuclei around the iron (Huijse et al., 2014).

The SN 1987A star used to have a mass of around 18 solar masses ahead of the explosion. A 1.5 solar mass center was located in the middle. The iron nuclei had become so warm that helium nuclei were removed. Temperatures and densities increased dramatically, and neutrinos were emitted at an astounding pace. A neutron star was generated when the gravitational pull grew too powerful. The nuclear powerful force gradually proved efficient when the neutrons drew near along, causing them to draw increasingly closer around each other. Whenever the neutrons were so near to each other that the nuclear force turned repellent, preventing the nucleons from merging, did this tendency cease. The material plunging inward during the gravitational collapse screeched to a stop and rebounded as a result of the powerfully repelling nuclear force (Boyd and Townes, 1977).

Figure 6.10. Photos were taken after the supernova of 1987 (on the left) and before it (on the right). The Anglo-Australian Telescope in Australia captured this blast. It was the first time a progenitor star had been discovered when the star detonated.

Source: https://www.britannica.com/place/Milky-Way-Galaxy.

The response of the internal structure halting and the outer layers continuing falling due to gravity caused a shock wave to go outward. The shock wave, on the other hand, was unable to break through the bulk of around 18 solar masses. The thick shock wave delayed the massive stream of neutrinos, that would ordinarily pass directly into the matter. The energy stored from the neutrinos was enough to provide the shock pulse the push it necessary to keep smashing through the star's outermost parts.

Neutrino generation is a vital sign of a supernova explosion since it demands a high neutrino rate. 3 hours when the first light hit Earth, neutrinos were discovered in the United States and Japan. Neutrinos may travel faster through the star versus light waves, which takes a more tortuous route. Neutrinos may exit from the center because the interaction energy is low sufficiently. The neutrinos' journey time to Earth as well as their energy spectrum matched estimates based on gravity and neutron star creation. With the solitary occurrence of supernova 1987A, our knowledge of star development advanced dramatically, and the science of neutrino astronomy got widely recognized. The observation and computations of SN 1987A yielded valuable information, along with an upper mass limit for the electron neutrino of 16 eV/c 2. Computer models have been fairly effective in explaining the ultraviolet outburst 2 hours just after the blast, the week-long spike in luminosity, and the subsequent decline in brightness. The discovery of SN 1987A provided us assurance that supernova observations may be utilized to estimate cosmic distances. SN 1987A happened in the Large Magellanic Cloud, was one of our nearest galaxies, and estimations of the range between this galaxy and Earth purely based on SN 1987A were all within 10% of our best prior predictions (160,000 ly). Supernova 1987A has served as a true proving ground for supernova ideas and simulations. A complicated system of encircling rings has been lighted up by radiation exposure (see Figure 6.10). Astrophysicists are attempting to recreate the formation of these rings.

6.9.4. Problems with the Big Bang

The Big Bang model was the preferred theory for the creation of the universe by the 1980s. Nonetheless, the simplistic Big Bang model proved unable to address at minimum three unsolved issues. They were:

- What causes the cosmos to be flat? According to Einstein's relativity theory, space must curve owing to gravitational pull. Space bends onto itself based on the quantity of material per unit of volume in the world (mass density), thus straight lines converge over a specific density termed the crucial concentration. Line segments split when mass density falls below the exact solution, and the cosmos extends indefinitely. An open universe is what it's named. The growth of the cosmos will ultimately come to a stop and afterward crash if the mass density exceeds a critical density. A sealed universe is what this is known as. In the 1980s, it looked like the universe was on the verge of becoming flat, or very close to being flat. Such an occurrence would be a once-in-a-lifetime event.
- Why would the cosmos seem to be so uniform in appearance no matter which way we appear? This is known as the horizon issue. If the world is 13.7 billion years old, the known universe's opposing ends are 27 billion light-years distant. Could these two places have identical microwave radiation? No matter which way we observe, the universe's temperature mirrors the 3 K microwave

baseline. How then can two parts of the cosmos with comparable temperatures and distances of 1019 m be unable to interact?

- Why haven't we found magnetic monopoles so far? Magnetic monopoles provide stability to Maxwell's electromagnetic equations, as well as satisfying other physics concepts.

6.9.5. The Inflationary Universe

In 1981, Alan Guth presented a modification to the traditional Big Bang concept, which substantially alleviated these problems. Guth suggested that among about 10^{-35} s and 10^{-31} s since the Big Bang, the universe grew by a ratio of 1050. Because of the separation of the nuclear and electroweak forces, this period is known as the inflationary era. Remembering that the cosmos was extremely tiny at the time, the enormity of this expansion is astounding. It's as though the hydrogen atoms electron, which is typically just 10^{-10} m from the proton, finally discovered itself 1,024 light-years distant! Following the inflationary epoch, the universe's development restarted per the traditional Big Bang scenario. Despite how difficult it may be to accept; inflation does mostly alleviate the three issues outlined. According to the inflationary hypothesis, the mass density must be very near to the exact solution. This overcomes the issue of flatness. Guth believed that the cosmos was so little that it had attained equilibrium before the occurrence of inflation; this describes why the world is homogenous. Cosmologists have added several proposals to Guth's inflationary theory since his work.

REFERENCES

1. Addison, G. E., Watts, D. J., Bennett, C. L., Halpern, M., Hinshaw, G., & Weiland, J. L., (2018). Elucidating ΛCDM: Impact of baryon acoustic oscillation measurements on the Hubble constant discrepancy. *The Astrophysical Journal, 853*(2), 119.
2. Appenzeller, I., (1967). A new polarimeter for faint astronomical objects. *Publications of the Astronomical Society of the Pacific,* 136–139.
3. Ashai, M., Mukherjee, R. G., Mundharikar, S. P., Kuanr, V. D., & Harikrishnan, R., (2022). Classification of astronomical objects using KNN algorithm. In: *Smart Intelligent Computing and Applications* (Vol. 1, pp. 377–387). Springer, Singapore.
4. Axelrod, T., Kantor, J., Lupton, R. H., & Pierfederici, F., (2010). An open source application framework for astronomical imaging pipelines. In: *Software and Cyberinfrastructure for Astronomy* (Vol. 7740, p. 774015). International Society for Optics and Photonics.
5. Basilakos, S., Nesseris, S., Anagnostopoulos, F. K., & Saridakis, E. N., (2018). Updated constraints on f (T) models using direct and indirect measurements of the Hubble parameter. *Journal of Cosmology and Astroparticle Physics, 2018*(08), 008.
6. Bates, R. H. T., & Cady, F. M., (1980). Towards true imaging by wideband speckle interferometry. *Optics Communications, 32*(3), 365–369.
7. Baxter, E. J., & Sherwin, B. D., (2021). Determining the Hubble constant without the sound horizon scale: Measurements from CMB lensing. *Monthly Notices of the Royal Astronomical Society, 501*(2), 1823–1835.
8. Baxter, J., (1989). Children's understanding of familiar astronomical events. *International Journal of Science Education, 11*(5), 502–513.
9. Bechhoefer, J., & Chabrier, G., (1993). On the fate of stars in high spatial dimensions. *American Journal of Physics, 61*(5), 460–462.
10. Berkhout, G. C., & Beijersbergen, M. W., (2008). Method for probing the orbital angular momentum of optical vortices in electromagnetic waves from astronomical objects. *Physical Review Letters, 101*(10), 100801.

11. Beskin, V. S., (2007). School of modern astrophysics. *Physics-Uspekhi, 50*(1), 103–105.
12. Birrer, S., & Treu, T., (2020). *TDCOSMO V: Strategies for Precise and Accurate Measurements of the Hubble Constant with Strong Lensing.* arXiv preprint arXiv:2008.06157.
13. Boesgaard, A. M., & Steigman, G., (1985). Big bang nucleosynthesis: Theories and observations. *Annual Review of Astronomy and Astrophysics, 23*(1), 319–378.
14. Bolejko, K., (2018). Emerging spatial curvature can resolve the tension between high-redshift CMB and low-redshift distance ladder measurements of the Hubble constant. *Physical Review D, 97*(10), 103529.
15. Bonnarel, F., Fernique, P., Bienaymé, O., Egret, D., Genova, F., Louys, M., & Bartlett, J. G., (2000). The ALADIN interactive sky atlas-a reference tool for identification of astronomical sources. *Astronomy and Astrophysics Supplement Series, 143*(1), 33–40.
16. Boyd, R. W., & Townes, C. H., (1977). An infrared upconverter for astronomical imaging. *Applied Physics Letters, 31*(7), 440–442.
17. Bruenn, S. W., & Mezzacappa, A., (1998). *Cosmic Closure: Relating the Ultimate Fate of Massive Stars and the Ultimate Fate of the Universe (No. ORNL/CP-100732).* Oak Ridge National Lab.(ORNL), Oak Ridge, TN (United States); Oak Ridge, TN.
18. Cabezas, C., Nakajima, M., Chang, C. H., Agúndez, M., Endo, Y., & Cernicharo, J., (2022). Laboratory observation and astronomical search of 1-cyano propargyl radical, HCCCHCN. *Astronomy & Astrophysics, 657*, A24.
19. Camenzind, M., (2007). *Compact Objects in Astrophysics* (pp. 1–25). Springer Berlin Heidelberg.
20. Carrasco-Davis, R., Cabrera-Vives, G., Förster, F., Estévez, P. A., Huijse, P., Protopapas, P., & Donoso, C., (2019). Deep learning for image sequence classification of astronomical events. *Publications of the Astronomical Society of the Pacific, 131*(1004), 108006.
21. Carroll, B. W., & Ostlie, D. A., (2006). An introduction to modern astrophysics and cosmology. *An Introduction to Modern Astrophysics and Cosmology/BW Carroll and DA Ostlie.* (2nd edn., pp. 1–20). *San Francisco: Pearson.*

22. Chen, G., Gott, J. R., & Ratra, B., (2003). Non-gaussian error distribution of Hubble constant measurements. *Publications of the Astronomical Society of the Pacific, 115*(813), 1269.
23. Chen, Y., Geng, C. Q., Cao, S., Huang, Y. M., & Zhu, Z. H., (2015). Constraints on a ϕCDM model from strong gravitational lensing and updated Hubble parameter measurements. *Journal of Cosmology and Astroparticle Physics, 2015*(02), 010.
24. Chen, Y., Kumar, S., & Ratra, B., (2017). Determining the Hubble constant from Hubble parameter measurements. *The Astrophysical Journal, 835*(1), 86.
25. Chopra, Y., Prabhala, N., & Tantri, P. L., (2017). Bank accounts for the unbanked: Evidence from a big bang experiment. *Robert H. Smith School Research Paper No. RHS, 2919091*.
26. Cyburt, R. H., Fields, B. D., Olive, K. A., & Yeh, T. H., (2016). Big bang nucleosynthesis: Present status. *Reviews of Modern Physics, 88*(1), 015004.
27. Davis, T. M., Hinton, S. R., Howlett, C., & Calcino, J., (2019). Can redshift errors bias measurements of the Hubble constant?. *Monthly Notices of the Royal Astronomical Society, 490*(2), 2948–2957.
28. Dayton, D., Gonglewski, J., Restaino, S., Martin, J., Phillips, J., Hartman, M., & Thiel, D., (2002). Demonstration of new technology MEMS and liquid crystal adaptive optics on bright astronomical objects and satellites. *Optics Express, 10*(25), 1508–1519.
29. De Bernardis, P., Ade, P. A., Bock, J. J., Bond, J. R., Borrill, J., Boscaleri, A., & Vittorio, N., (2000). A flat Universe from high-resolution maps of the cosmic microwave background radiation. *Nature, 404*(6781), 955–959.
30. DeFond, M. L., (2019). Why Accounting matters: Evidence from accounting's "big bang". *Journal of International Accounting Research, 18*(3), 87–96.
31. Deligny, O., & Suomijaervi, T., (2015). High energy cosmic rays: A riddle of modern astrophysics. *Reflets de la Physique,* 31–37.
32. Farooq, O., Crandall, S., & Ratra, B., (2013). Binned Hubble parameter measurements and the cosmological deceleration–acceleration transition. *Physics Letters B, 726*(1–3), 72–82.

33. Feduccia, A., (2003). 'Big bang' for tertiary birds?. *Trends in Ecology & Evolution, 18*(4), 172–176.

34. Flin, P., & Duerbeck, H. W., (1989). Morphological cosmology. *Morphological Cosmology, 332*.

35. Freedman, W. L., (2021). Measurements of the Hubble constant: Tensions in perspective. *The Astrophysical Journal, 919*(1), 16.

36. Gawiser, E., & Silk, J., (2000). The cosmic microwave background radiation. *Physics Reports, 333*, 245–267.

37. Gurzadyan, V. G., & Penrose, R., (2010). *Concentric Circles in WMAP Data May Provide Evidence of Violent Pre-Big-Bang Activity*. arXiv preprint arXiv:1011.3706.

38. Gustafsson, B., Edvardsson, B., Eriksson, K., Jørgensen, U. G., Nordlund, Å., & Plez, B., (2008). A grid of MARCS model atmospheres for late-type stars-I. Methods and general properties. *Astronomy & Astrophysics, 486*(3), 951–970.

39. He, S., Li, Y., Feng, Y., Ho, S., Ravanbakhsh, S., Chen, W., & Póczos, B., (2019). Learning to predict the cosmological structure formation. *Proceedings of the National Academy of Sciences, 116*(28), 13825–13832.

40. Hogan, C. J., (1994). *Giant Branch Mixing and the Ultimate Fate of Primordial Deuterium in the Galaxy.* arXiv preprint astro-ph/9407038.

41. Hoyle, F., & Fowler, W. A., (1960). Nucleosynthesis in Supernovae. *The Astrophysical Journal, 132*, 565.

42. Hubble, S. M., Kyte, H. L., Gooding, K., & Shore, A. C., (2009). Variability in sublingual microvessel density and flow measurements in healthy volunteers. *Microcirculation, 16*(2), 183–191.

43. Hughes, J. P., Rakowski, C. E., Burrows, D. N., & Slane, P. O., (1999). Nucleosynthesis and mixing in Cassiopeia A. *The Astrophysical Journal, 528*(2), L109.

44. Huijse, P., Estevez, P. A., Protopapas, P., Principe, J. C., & Zegers, P., (2014). Computational intelligence challenges and applications on large-scale astronomical time series databases. *IEEE Computational Intelligence Magazine, 9*(3), 27–39.

45. Hujeirat, A. A., (2018). On the ultimate fate of massive neutron stars in an ever expanding universe. *Journal of Modern Physics, 9*(01), 51.

46. Islam, J. N., (1977). Possible ultimate fate of the universe. *Quarterly Journal of the Royal Astronomical Society, 18*, 3.

47. Izotov, Y. I., & Thuan, T. X., (2010). The primordial abundance of 4He: Evidence for non-standard big bang nucleosynthesis. *The Astrophysical Journal Letters, 710*(1), L67.
48. Jackson, N., (2015). The Hubble constant. *Living Reviews in Relativity, 18*(1), 1–51.
49. Kardashev, N. S., (1997). Cosmology and civilizations. *Astrophysics and Space Science, 252*(1), 25–40.
50. Kawasaki, M., Kohri, K., Moroi, T., & Yotsuyanagi, A., (2008). Big-bang nucleosynthesis and gravitinos. *Physical Review D, 78*(6), 065011.
51. Kirsebom, O. S., Cederkall, J., Jenkins, D. G., Joshi, P., Julin, R., Kankainen, A., & Tengblad, O., (2016). Electron capture on "2" 0Ne and the ultimate fate of stars in the mass range 8-10 M⊙. In: *Proceedings of the DAE-BRNS Symposium on Nuclear Physics* (Vol. 61).
52. Kramar, N., & Ilchenko, O., (2021). From intriguing to misleading: The ambivalent role of metaphor in modern astrophysical and cosmological terminology. *Amazonia Investiga, 10*(46), 92–100.
53. Kurt, V. G., (1997). Cosmic gamma bursts-phantoms in modern astrophysics. *Priroda,* 74.
54. Lang, D., Hogg, D. W., & Mykytyn, D., (2016). *The Tractor: Probabilistic Astronomical Source Detection and Measurement.* Astrophysics Source Code Library, ascl-1604.
55. Larson, S. L., (2005). LISA: A modern astrophysical observatory. *Proc. of the 33rd SLAC Summer Institute on Particle Physics (SSI 2005): Gravity in the Quantum World and the Cosmos.* Menlo Park, USA: Stanford University eConf.
56. Lemoine, M., & Sigl, G., (2001). *Physics and Astrophysics of Ultra High Energy Cosmic Rays* (Vol. 576). Springer Science & Business Media.
57. Maeder, A., & Lequeux, J., (1982). The ultimate fate of wolf-rayet stars as supernovae. *Astronomy and Astrophysics, 114*, 409–413.
58. Malville, J. M., (1991). Astrophysics, cosmology, and the interior space of Indian myths and temples. *Concepts of Space, Ancient and Modern,* 123–144.
59. Meyer, F., & Meyer-Hofmeister, E., (1999). *On the Ultimate Fate of AM Her Stars*. arXiv preprint astro-ph/9905068.

60. Michelson, A. A., (1920). On the application of interference methods to astronomical measurements. *The Astrophysical Journal, 51*, 257.
61. Momcheva, I. G., Brammer, G. B., Van, D. P. G., Skelton, R. E., Whitaker, K. E., Nelson, E. J., & Wuyts, S., (2016). The 3D-HST survey: Hubble space telescope WFC3/G141 grism spectra, redshifts, and emission line measurements for~ 100,000 galaxies. *The Astrophysical Journal Supplement Series, 225*(2), 27.
62. Morrison, P., (1958). On gamma-ray astronomy. *Il Nuovo Cimento (1955–1965), 7*(6), 858–865.
63. Morse, J. A., Kellogg, J. R., Bally, J., Davidson, K., Balick, B., & Ebbets, D., (2001). Hubble space telescope proper-motion measurements of the η carinae nebula. *The Astrophysical Journal, 548*(2), L207.
64. Mukherjee, S., Lavaux, G., Bouchet, F. R., Jasche, J., Wandelt, B. D., Nissanke, S., & Hotokezaka, K., (2021). Velocity correction for Hubble constant measurements from standard sirens. *Astronomy & Astrophysics, 646*, A65.
65. Ochsenbein, F., Bauer, P., & Marcout, J., (2000). The VizieR database of astronomical catalogues. *Astronomy and Astrophysics Supplement Series, 143*(1), 23–32.
66. Partridge, R. B., & Partridge, R. B., (1995). *3K: The Cosmic Microwave Background Radiation* (Vol. 393). Cambridge: Cambridge University Press.
67. Philcox, O. H., Sherwin, B. D., Farren, G. S., & Baxter, E. J., (2021). Determining the Hubble constant without the sound horizon: Measurements from galaxy surveys. *Physical Review D, 103*(2), 023538.
68. Pino, T., Dartois, E., Cao, A. T., Carpentier, Y., Chamaillé, T., Vasquez, R., & Bréchignac, P., (2008). The 6.2 band position in laboratory and astrophysical spectra: A tracer of the aliphatic to aromatic evolution of interstellar carbonaceous dust. *Astronomy & Astrophysics, 490*(2), 665–672.
69. Ringwald, A., (2005). *How to Detect Big Bang Relic Neutrinos?*. arXiv preprint hep-ph/0505024.
70. Robertson, H. P., (1933). Relativistic cosmology. *Reviews of Modern Physics, 5*(1), 62.
71. Schlegel, D. J., Finkbeiner, D. P., & Davis, M., (1998). Maps of dust infrared emission for use in estimation of reddening and cosmic

microwave background radiation foregrounds. *The Astrophysical Journal, 500*(2), 525.

72. Schramm, D. N., & Turner, M. S., (1998). Big-bang nucleosynthesis enters the precision era. *Reviews of Modern Physics, 70*(1), 303.
73. Sergii, R., Yaroslav, S., & Vadym, T., (2018). Cosmology in the philosophical education of Ukraine: History and modern condition. *Философия и Космология, 20*, 128–138.
74. Shi, X., & Turner, M. S., (1998). Expectations for the difference between local and global measurements of the Hubble constant. *The Astrophysical Journal, 493*(2), 519.
75. Shore, S. N., (2002). *The Tapestry of Modern Astrophysics*, 888.
76. Smith, H. A., (2017). Questioning Copernican mediocrity: Modern astrophysics can intimate our cosmic significance. *American Scientist, 105*(4), 232–240.
77. Smith, N., & Owocki, S. P., (2006). On the role of continuum-driven eruptions in the evolution of very massive stars and population III stars. *The Astrophysical Journal, 645*(1), L45.
78. Somov, B. V., (2012). *Fundamentals of Cosmic Electrodynamics* (Vol. 191). Springer Science & Business Media.
79. Srianand, R., Petitjean, P., & Ledoux, C., (2000). The cosmic microwave background radiation temperature at a redshift of 2.34. *Nature, 408*(6815), 931–935.
80. Starck, J. L., Murtagh, F., & Louys, M., (1995). Astronomical image compression using the pyramidal median transform. In: *Astronomical Data Analysis Software and Systems IV* (Vol. 77, p. 268).
81. Sulentic, J. W., & Tifft, W. G., (1979). *The Revised New General Catalogue of Nonstellar Astronomical Objects*, 23–37.
82. Tacchella, S., Carollo, C. M., Renzini, A., Schreiber, N. F., Lang, P., Wuyts, S., & Zamorani, G., (2015). Evidence for mature bulges and an inside-out quenching phase 3 billion years after the big bang. *Science, 348*(6232), 314–317.
83. Tanndal, J., & Waldenström, D., (2018). Does financial deregulation boost top incomes? Evidence from the big bang. *Economica, 85*(338), 232–265.
84. Tilvi, V., Malhotra, S., Rhoads, J. E., Coughlin, A., Zheng, Z., Finkelstein, S. L., & Yang, H., (2020). Onset of cosmic reionization:

Evidence of an ionized bubble merely 680 Myr after the big bang. *The Astrophysical Journal Letters, 891*(1), L10.

85. Tsallis, C., Barreto, F. S., & Loh, E. D., (1995). Generalization of the Planck radiation law and application to the cosmic microwave background radiation. *Physical Review E, 52*(2), 1447.
86. Wagoner, R. V., (1973). Big-bang nucleosynthesis revisited. *The Astrophysical Journal, 179*, 343–360.
87. Walker, T. P., Steigman, G., Schramm, D. N., Olive, K. A., & Kang, H. S., (1991). Primordial nucleosynthesis redux. *The Astrophysical Journal, 376*, 51–69.
88. Webster, A., (1991). Comparison of a calculated spectrum of C60H60 with the unidentified astronomical infrared emission features. *Nature, 352*(6334), 412–414.
89. Weigelt, G. P., (1977). Modified astronomical speckle interferometry "speckle masking". *Optics Communications, 21*(1), 55–59.
90. Wenger, M., Ochsenbein, F., Egret, D., Dubois, P., Bonnarel, F., Borde, S., & Monier, R., (2000). The SIMBAD astronomical database-The CDS reference database for astronomical objects. *Astronomy and Astrophysics Supplement Series, 143*(1), 9–22.
91. White, M., Scott, D., & Silk, J., (1994). Anisotropies in the cosmic microwave background. *Annual Review of Astronomy and Astrophysics, 32*(1), 319–370.
92. Wilson, R. W., (1979). The cosmic microwave background radiation. *Science, 205*(4409), 866–874.
93. Wright, E. L., Meyer, S. S., Bennett, C. L., Boggess, N. W., Cheng, E. S., Hauser, M. G., & Wilkinson, D. T., (1992). Interpretation of the cosmic microwave background radiation anisotropy detected by the COBE differential microwave radiometer. *The Astrophysical Journal, 396*, L13–L18.
94. Wu, H. Y., & Huterer, D., (2017). Sample variance in the local measurements of the Hubble constant. *Monthly Notices of the Royal Astronomical Society, 471*(4), 4946–4955.
95. Wu, W. K., Motloch, P., Hu, W., & Raveri, M., (2020). Hubble constant difference between CMB lensing and BAO measurements. *Physical Review D, 102*(2), 023510.
96. Yurchenko, S. N., Tennyson, J., Bailey, J., Hollis, M. D., & Tinetti, G., (2014). Spectrum of hot methane in astronomical objects using

a comprehensive computed line list. *Proceedings of the National Academy of Sciences, 111*(26), 9379–9383.

97. Zhang, Y., & Zhao, Y., (2004). Automated clustering algorithms for classification of astronomical objects. *Astronomy & Astrophysics, 422*(3), 1113–1121.
98. Zhao, Y., & Zhang, Y., (2008). Comparison of decision tree methods for finding active objects. *Advances in Space Research, 41*(12), 1955–1959.

CHAPTER 7

PHYSICS OF SEMICONDUCTING LASERS

CONTENTS

7.1. INTRODUCTION

A laser is an optical instrument that produces and magnifies a coherent, homogeneous beam of light that is both small and strong. (Gihleengen et al., 2007) LASER is an acronym for "Light Amplification by Stimulated Emission of Radiation."

In 1958, Arthur L. Schawlow and Charles H. Townsey published the science article Infrared and Optical Masers in the American Physical Society's Physical Review (Anscombe, 2007). This article marks the birth of the laser. Even so, Albert Einstein was the only one to suggest the theory of "Stimulated Emission" in 1917 (Basov et al., 1961), the procedure upon which laser beams are centered (Figure 7.1).

Figure 7.1. Arthur L. Schawlow with a ruby laser, 1961.

Source: http://texts.cdlib.org/view?docId=kt5b69n7k2&brand=calisphere&doc.view=entire_text.

Semiconductors can be used as small, highly efficient photon sources in lasers. These are called semiconductor lasers. In 1961, Basov et al. proposed that photon emission might be generated in semiconductors by the replication of transmitters implanted along a p-n junction. This was the beginning of semiconductor lasers. In 1962, the initial p-n junction lasers were constructed in GaAs (infrared) (Hall et al., 1962) as well as GaAsP (visible) (Hawkes and Latimer, 1995); this elevated the laser's significance in semiconductor gadget technology (Lan and Ding, 2012).

Presently, lasers are utilized in numerous fields, including information exchange, related industries, pharmaceutics, environmental conservation, and investigations. The laser is now one of the greatest effective methods for physicists, chemists, biologists, and physicians worldwide (Crowley et al., 2012). Due to its low cost and diminutive size, the semiconductor laser is becoming particularly appealing. In comparison to the vast majority of other laser kinds, it is super effective and contains very simple power supply specifications (Ready, 1997).

7.2. BASIC LASER THEORY

7.2.1. Stimulated Emission

Photons and atomic energy stages may be used to describe the fundamentals of a laser. We'll use an atomic electron with just energy states divided by an energy E for clarity. Whenever the system would be in a lower energy level, it may well be stimulated to a higher energy level by absorbing light with energy $E = h\nu$, wherein; h is Planck's constant and f is the photon's frequency. Eventually, the identical atom would go to the bottom level and produce a photon with a similar frequency as the one it received. This is known as spontaneous emission, and it results in light in a random way and amplitude (Paschotta et al., 2020).

The method of spontaneous emission is used in lasers. The existence of photons of the appropriate wavelength induces an electron in a greater or activated form to descend to a low energy state in this procedure. A photon is emitted as a consequence of this decrease. The activating rays have similar energy as the distinction between the two energy states, which is h. As a result, because all photons have a similar energy, the emission will be homogenous. It will be coherent as well since all photons produced will be in sequence and reinforce each other. Figure 7.2 (Schubert, 2006) illustrates the procedure.

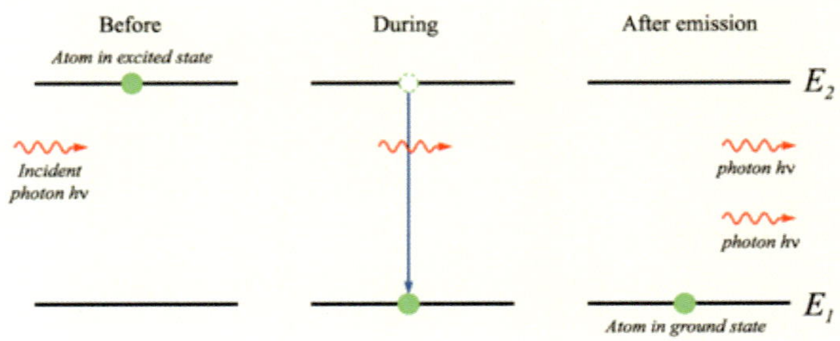

Figure 7.2. Process of stimulated emission.

Source: https://en.wikipedia.org/wiki/Stimulated_emission.

7.2.2. Population Inversion

Consider a group of atoms with two energy levels, E_1 and E_2, for every atom. The relative numbers of atoms in every stage are N_1 and N_2. When passing a beam of light with frequency $v_{12} = (E_2\ E_1)/h$ across this group of atoms, either absorb or spontaneous emission procedures happen. The likelihood of a photon getting captured by an atom at a lower energy level is identical to the chance of the photon generating triggered emission whenever it meets an atom in an excited state. This indicates that photons in the beam would be destroyed by absorbance at a speed comparable to N_1 and generated via spontaneous emission at a speed comparable to N_2. The time-dependent variation in the amount of photons n describes according to:

$$\frac{dn}{dt} = C(N_2 - N_1) \qquad (1)$$

where; *C* is a constant.

When N_2 is more than N_1, the amount of photons will increase and laser light will be created. This phenomenon is known as population inversion. Sadly, whenever atoms are in thermal equilibrium, N_1 is always bigger than N_2, and outer power input, also known as 'pumping,' is required to accomplish population inversion (Samuel and Turnbull, 2004; Grundmann, 2010).

Imagine a three-level laser as an instance. It comprises a rod-shaped ruby rise media and a high-powered flash tube that provides light for pumping. Ruby is essentially an Al_2O_3 crystal with a trace quantity of Cr.

It is the chromium's excitement rates that produce the phases required for lasing. Figure 7.3 is a diagrammatic representation of the three lasering levels (Samuel and Turnbull, 2007). Levels 1 and two are so high beyond the lowest energy so, in thermal equilibrium, their numbers are small when compared to the ground stage.

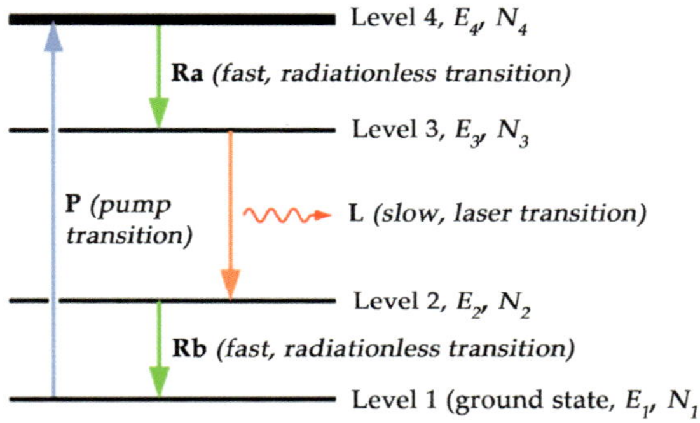

Figure 7.3. A schematic diagram of a three-level laser showing the pumping and laser transitions.

Source: http://electrons.wikidot.com/principle-and-application-of-laser.

The pumping illumination comprises the frequency part necessary to stimulate the atoms from zero level to 2nd level. Level 2 atoms undergo a fast (nonradiative) shift to level 1, which contains a relatively long lifespan. Atoms begin to collect on level 1, and with powerful enough pumping, a population inversion occurs among levels 1 and 0. Every light with the proper frequency will trigger emission among these two levels, producing laser light (Yamada, 2014).

7.2.3. Physical Structure

A laser is typically made up of a nonlinear material with reflectors on both sides (see Figure 7.4) and a few types of external sources of energy for pushing. The photon stream will magnify as it passes through the transmission medium, which may be a solid, liquid, or gas. The beam intensity would not be exceptionally powerful after only one passage of the laser source except if the pumping is exceedingly vigorous or the gain medium is lengthy. To produce optical feedback, mirrors are utilized at both ends. Every photon passes through the medium many times as a result, causing additional emissions until exiting (Balkanski et al., 2000).

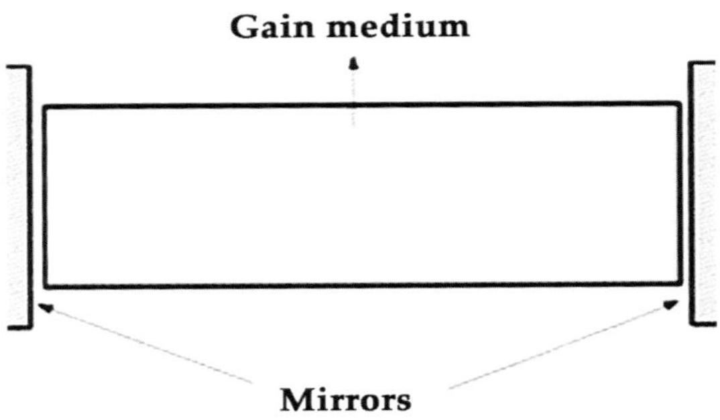

Figure 7.4. A laser with a gain medium and mirrors on each side to provide optical feedback.

Source: https://www.premiumtattooremoval.com/laser-physics.

The gap among the mirrors is known as the laser cavity (also known as the optical cavity). One mirror is extremely reflective, and the other has a reflectivity of roughly 95%, enabling the laser light to pass across. The period of the wavefront at any given position should be similar (to about a digit multiple of 2), that is one of the conditions for steady behavior for a plane wavefront traveling to and fro inside the hole. As a result, the laser is also called a laser oscillator (Stern and Woodall, 1974).

Laser pumping may be accomplished in a variety of ways, including optical, electrical, and chemical approaches. To produce sufficient excited atoms to outweigh those lost due to stimulated emission and other causes, the energy source's brightness must be greater than a particular threshold value. The laser beam will not increase forever, even with strong sufficient pumping. Since the population inversion decreases as the cavity's irradiance rise, the gain tends to diminish. Cavity loss is unavoidable owing to absorption as well as optical components that link light out the cavity. Whenever the losses per travel match the gains per transit, the irradiance achieves saturation (Alferov, 2001; Yang, 2005).

7.3. SEMICONDUCTOR LASERS

7.3.1. Basic Physics

The gain medium of a semiconductor laser comprises sheets of the doped semiconducting medium. As seen in Figure 7.5, light is released at an interface among two areas of opposing doping. The manner of functioning differs slightly from the laser discussed in the preceding chapter. Rather than activating atomic electrons at various sites in a material by flashing light, the vast amount of electrons at a joint cause the phase difference. In addition, pumping is accomplished by electrons (injecting carriers). As a result of the crystalline structure of semiconductors, the electrons do not contain discrete energy states like the chromium ions in ruby, but rather continual ranges of possible states. We are solely interested in the highest filled valence band Ev as well as the (intrinsic) vacant conduction band Ec beyond it. The lasing activity will occur among the margins of such two bands; hence, the frequency of the produced light will be equal to the bandgap E_g (Hamaguchi and Hamaguchi, 2010).

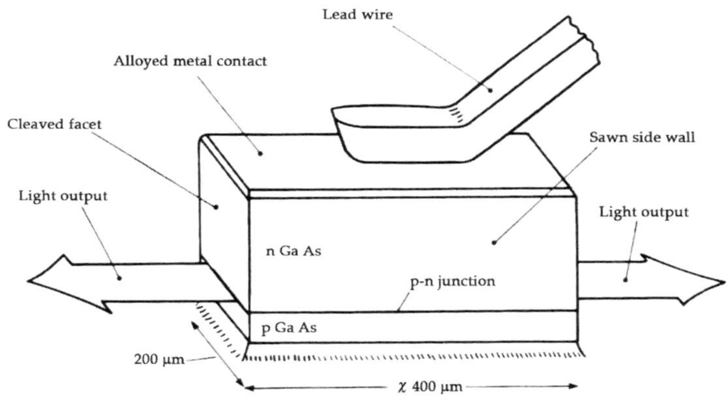

Figure 7.5. A simple p-n junction laser.

Source: https://www.electronicshub.org/laser-diode-working-structure-types-uses.

Figure 7.5 depicts the connection between the energy E and the wave vector k (momentum) for electrons in a crystal that is symbolized by a propagating plane wave. A straight band gap indicates that its minimum electron energy in the valence band and the greatest electron energy in the conductive band have a similar k. This implies as an electron can jump from Ec to Ev (releasing a photon) without no affecting its momentum (Figure 7.6) (Wieman and Hollberg, 1991; Łukasiak and Jakubowski, 2010).

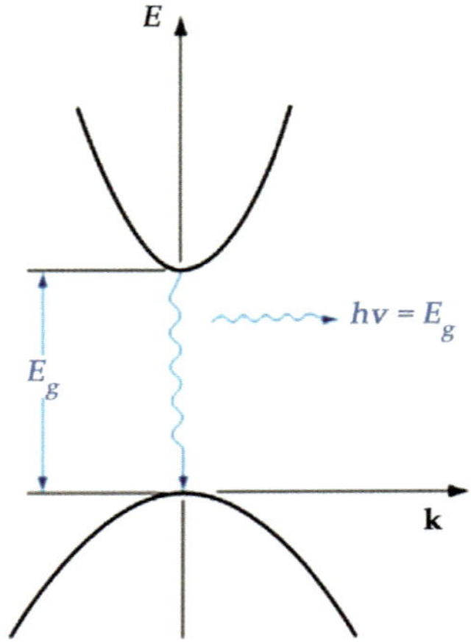

Figure 7.6. A direct transition accompanied by a photon emission.

Source: https://en.wikipedia.org/wiki/Photoelectric_effect.

Similar to other laser mediums, it is required to invert the community to produce a positive net gain. Figure 7.7 depicts the band chart of a p-n junction, which is the easiest approach to explaining population inversion in semiconductors. When either side's Fermi levels are in their corresponding bands (e.g., are severely doped) and the junction would be in forwarding bias, the Fermi states would divide into quasi-Fermi levels F_n and F_p instead of remaining as a single line across the center (Yablonovitch and Kane, 1988).

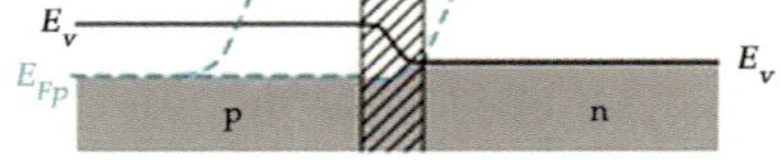

Figure 7.7. Band chart of a p-n junction laser within forwarding bias, as shown. Crosshatched area denotes the zone of inversion at the junction.

Source: http://www.industrial-electronics.com/solid-state-elec-dev_8b.html.

This implies that in a narrow area known as the inversion zone, a significant number of holes and electrons would occupy a similar space, enabling many recombination. Since there are a variety of filled stages in such bands, changes may occur with energies spanning from hv = (F_n F_p) to h = E_g, not just against the boundaries of the bands. The resonant cavity and the reality that recombination radiation at h = E_g is intense (Herd et al., 1997) define the dominant transitions.

7.3.2. Homojunction Laser

An n-doped slice of GaAs was the basic and oldest version of a semiconducting laser. Diffusion p-doped a section of the substance to form a p-n junction, as shown in Figure 7.8. The homostructure laser has been so named because it uses just one kind of substance. Two of the corners of a crystallographic plane are cleaved and the other two are roughened to create an optical cavity. By leaving the split sides untreated, Fresnel reflectivity is achieved around 30% due to the sheer considerable disparity in the refractive index.

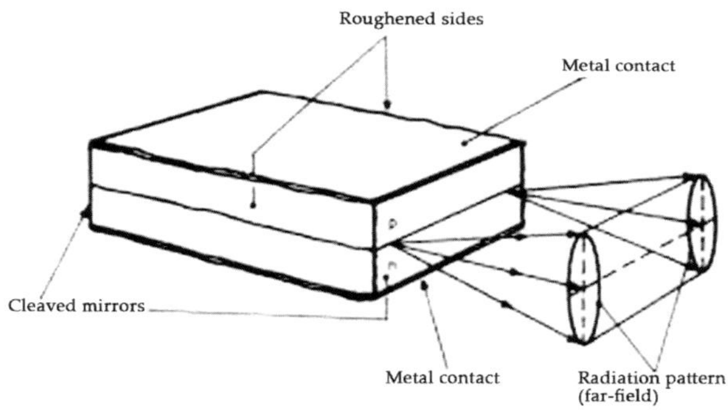

Figure 7.8. Diagram of a homostructure laser showing light emission from cleaved edge.

Source: https://www.researchgate.net/figure/Schematic-and-pumping-geometry-for-the-InGaN-GaN-single-heterostructure-SH_fig1_202846499.

The rough surface sides of such half mirrors will prohibit specular reflection, creating a Fabry-Perot optical cavity. A graded index of refraction will result from the variation in electron density throughout the junction, functioning as an optical waveguide. The density J_{th} required for this gadget to lase at 80 K is in the 500–1,000 A cm^2 range. J_{th}, on the other hand, is extremely sensitive to temperature, measuring between 30 and 50 k

Acm[2] at ambient temperature. The width of the inversion zone grows with temperature, requiring a larger volume of implanted particles, which is one of the major causes of the high cutoff point. This should result in a flatter refractive index distribution, decreasing the waveguiding capabilities (Deacon, 1981).

7.3.3. Heterojunction Laser

For laser systems, the necessity to preserve a tiny inversion zone and strong optical waveguiding qualities necessitate the inclusion of extra materials. By interspersing sheets of materials along with varying bandgaps, it is better to alter performance by restricting the propagation of carriers and confining the optical wave. As shown in the section on materials, the bandgap of numerous ternary and quaternary compounds may well be modified by altering the elemental composition. Utilizing substances with the same lattice constant eliminates defect situations at junctions (Wang, 2000).

The single heterojunction laser is the basic heterostructure. Similar to the homostructure, it consists of a GaAs active material in which the p-n junction is situated. As illustrated in Figure 7.8, a coating of (AlGa). As has been formed on over of the p-region. As seen in Figure 7.9, this coating will operate as a blocker for the injected particles due to its larger bandgap. This connection to a substance with a distinct band gap would likewise produce a sudden shift in refractive index, resulting in enhanced waveguiding capabilities (Manheimer et al., 2008).

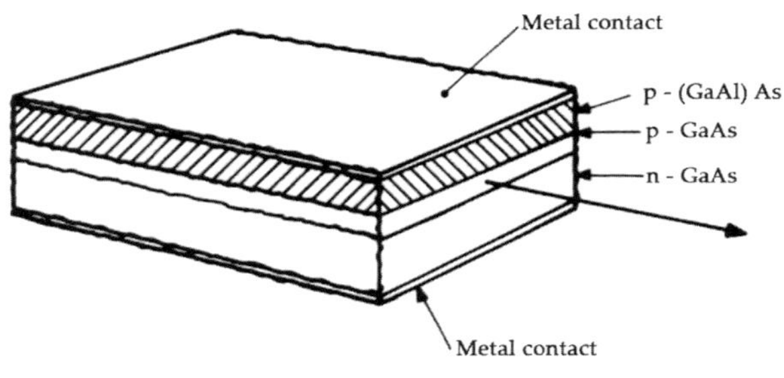

Figure 7.9. Diagram of a single heterostructure.

Source: https://www.researchgate.net/figure/Schematic-and-pumping-geometry-for-the-InGaN-GaN-single-heterostructure-SH_fig1_202846499.

This construction has the benefit of being simple to manufacture. On a substrate of n-type GaAs, P-type (AlGa) As might well be formed epitaxially.

By increasing the temperature, dopants would spread from the top layer into the substrate, forming a p-doped GaAs area. The threshold current density for this arrangement is around 6,000–8,000 Acm^2, which is 5–10 times less than that of a homostructure, according to Arimitsu and Shibata (1983); and Stenholm (1986).

On the double heterojunction, the cutoff current may be reduced even more. As illustrated in Figure 7.10, the active area of this structure is sandwiched among p- and n-type AlGaA. In this gadget, J_{th} might well be decreased such that continual-wave operation at room temperature is possible (Dowden, 2009).

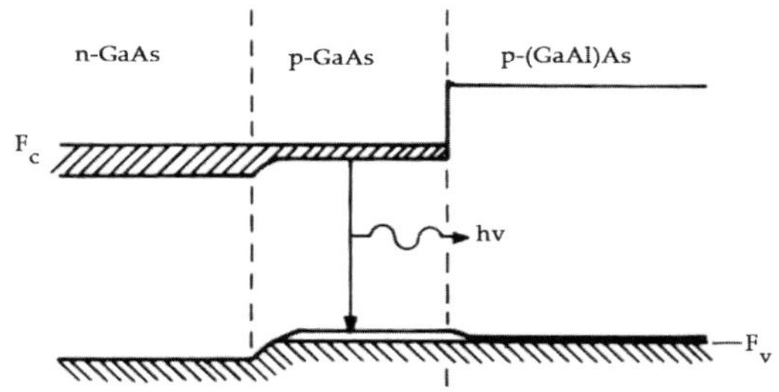

Figure 7.10. Distribution of conduction and valence bands perpendicular to a p-n junction in a single heterostructure laser.

Source: https://www.brainkart.com/article/Hetro-Junction-Laser--Principle,-Construction,-Working,-Characteristics,-Advantages,-Disadvantages-and-Applications_6887/.

By restricting the injected current to a thin band of the active layer, a more potent variation of the double heterojunction laser may be produced. This higher density over a short region will permit greater current reduction (Numai, 2015).

In all of the examined laser systems so far, particle confinement, as well as waveguiding, have relied on a similar junction to a substance with a distinct bandgap. It is more efficient to separate these functions so that they may be optimized independently. In a Separate Confinement Channel structure, as seen in Figure 7.11, carrier recombination occurs in a small area d, while optical waveguiding is limited in a larger region w. Again, examining the GaAs Al_xGa1_x As-system, carrier confinement requires a transition to a region having less Al than waveguiding (Agrawal and Dutta, 1986).

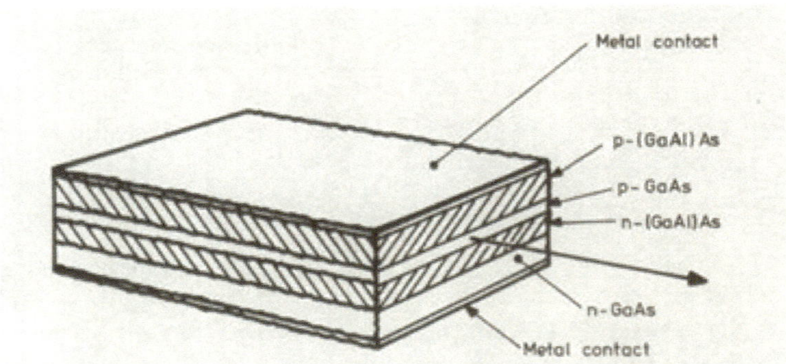

Figure 7.11. Diagram of a double heterostructure.

Source: https://www.researchgate.net/figure/Schematic-of-the-double-hetero-junction-LED-structure_fig2_4140572.

By progressively varying the Al concentrations on each side of the carrier confinement zone, as seen in Figure 7.12, waveguiding qualities comparable to those of a fiber optic route might well be produced (Coldren et al., 2004; Ohtsubo, 2012).

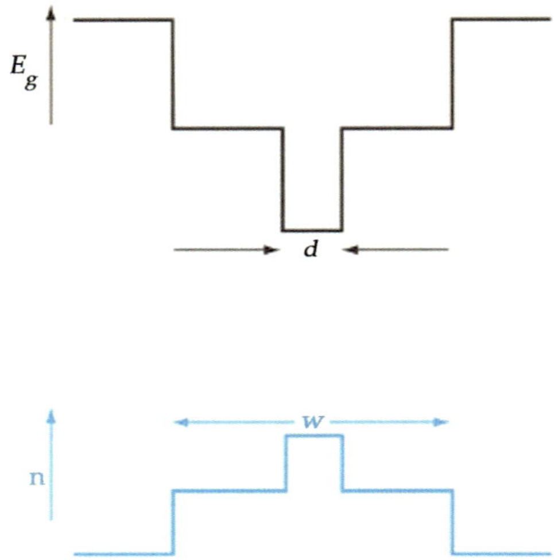

Figure 7.12. Separate changes in compounds will confine carriers to the thin region 'd,' and waveguiding will take place in a larger region 'w.'

Source: https://www.sciencedirect.com/topics/engineering/channel-waveguide.

7.4. SEMICONDUCTOR LASER MATERIALS

Several semiconductive substances have been used to create semiconductor lasers. The primary goal of studying the utilization of various substances is to increase the scope of wavelengths that may be used (Tucker, 1985).

The majority of semiconductor lasers are made up of substances from the periodic table's III and V columns (Figure 7.13). Gallium arsenide, for instance, was the first semiconductive laser material. The earliest semiconductor lasers were created from crystals with a p- and n-type gallium arsenide junction. These have already been surpassed by ternary or quaternary semiconductor materials, which comprise three or four components from the periodic table's columns III and V. The percentage of these chemicals (Henry, 1982).

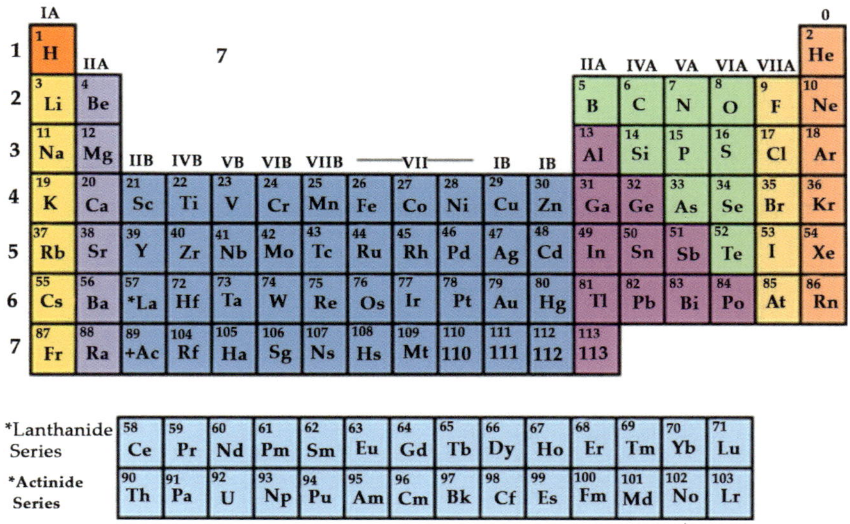

Figure 7.13. Periodic table of elements.

Source: https://www.britannica.com/science/periodic-table.

Various material qualities may be created by varying the size of every element. The ternary compound $Al_{1-x}Ga_xAs$ (aluminum gallium arsenide) and the quaternary complex $In_{1-x}Ga_xAs_{1-y}$ Py are two examples (indium gallium arsenide phosphide). x and y are composition variables with values ranging from 0 to 1 in this format. They provide the compound's specific makeup. As a result, the $Al_{1-x}Ga_xAs$ system may change frequently from AlAS (x=0) to GaAs (x=1) (Iga et al., 1988).

The alloy aluminum gallium arsenide is perhaps the most popular semiconductor laser source. In a variety of ternary solid solutions of III/V compounds, Figure 7.14 depicts the connection between the lattice parameter and the bandgap. This kind of graphic is excellent for demonstrating the variety of band gaps that may be formed whenever the content of a solid solution is changed with no changing the lattice constant (Ettenberg et al., 1979).

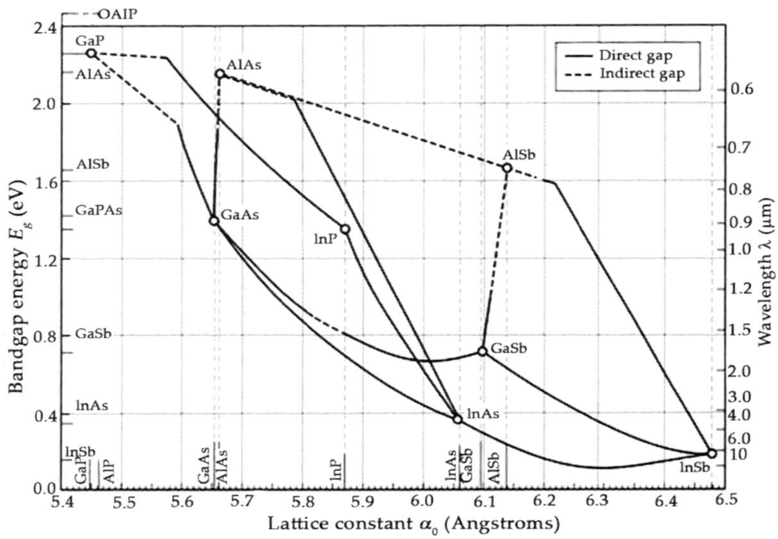

Figure 7.14. At ambient temperature, the bandgap energy and lattice constant of several III/V semiconductors.

Source: https://www.researchgate.net/figure/Bandgap-energy-versus-lattice-constant-of-various-III-V-semiconductors-at-room_fig2_241880293.

Different kinds of semiconductor lasers, such as substances from the periodic table's II and VI columns, are also known. These apps work in the far-infrared spectrum, and they are far more costly and scarce than III/V compounds (De Loach et al., 1973; Dumke, 1975).

Semiconductor laser components must be effective light producers. It must also be feasible to generate p-n junctions and, in most situations, heterojunction barriers. As a result, certain materials are just unsuitable for semiconductor applications. These materials include indirect bandgap semiconductors, which are insufficiently effective light emitters for effective laser manufacturing (Kroemer, 1963; Van Ruyven, 1984).

Although the II/VI compounds are typically excellent light emitters, forming connections is problematic. By using N as the acceptor, modern methods like molecular beam epitaxy (MBE) and metal-organic vapor phase epitaxy (MOVPE) may form connections in ZnS, ZnSe, ZnTe, and alloys of these materials. These materials can be utilized to create green and blue-green light-emitting lasers (Tucker, 1981; Eliseev, 1996).

7.5. APPLICATIONS

There are a variety of uses for lasers. Due to the quest for lower and more compact lasers, semiconductor lasers are highly desired. Highly versatile, they are employed in fields such as optical data transfer, optical computing, metrology, spectroscopy, materials science, pumping solid-state lasers, and a variety of medicinal therapies (Figure 7.15) (Childs et al., 1986; Gmachl et al., 1999).

Figure 7.15. Lasers in a DVD player.

Source: https://danyk.cz/laser3_en.html.

Most of us utilize lasers daily. For instance, semiconductor lasers are utilized to look at the information recorded on CDs and digital versatile discs (DVDs). The discs' memory capacity is negatively proportionate to the laser wavelength squared. The development of lasers with wavelengths shorter enables the storage of more data on discs. Moreover, a high spatial consistency is required for recording extremely tiny locations on the recording media (Hader et al., 1999; Kränkel et al., 2016).

In fiber-optic networks, semiconductor lasers are also employed as emitters. These are components of an optical fiber transmission medium that establish a data link between two locations. These sorts of optical cables are utilized between being an amplifier and the loudspeakers (Coleman, 2012).

Laser pointers are manufactured using low-cost semiconductor lasers. Many laser pointers are using a tiny GaInP/AlGaInP laser diode that emits red light. Laser pointers that operate in the green and blue spectral areas are more costly. Due to their compact size and ability to be operated from a distant location, laser pointers are useful for use in presentations, for instance. Additionally, they may be used for specific optical distance measurements (Figure 7.16) (Zhao et al., 2015; Adachi and Sandanayaka, 2020).

Figure 7.16. Laser pointers.

Source: https://turbofuture.com/consumer-electronics/Fun-Things-to-do-With-a-Laser-Pointer.

7.6. RECENT DEVELOPMENT: QUANTUM DOT LASERS

Almost all of the research in semiconductor laser technology since about the 1960s has indeed been aimed at improving the complex structure as well as implementing new materials. Many anticipate that the quantum dot laser may likely overtake this sort of semiconducting laser, particularly in the area of data chip transmission (Oyama et al., 2020).

A quantum dot laser is a semiconductor laser in which the effective gain medium is quantum dots in layers. Semiconductor nanocrystals are used to make the laser quantum dots. Quantum dots may range in size from 2 to 10 nm, although self-assembled quantum dots are generally around 10 and 50 nanometers, as seen in Figure 7.17. Quantum dots have a limited number of conduction electrons as well as valence band holes (also known as excitons) that they can restrict in all three directions. Quantum dots have an electrical structure comparable to atoms because of the dense packaging of electric charge. Quantum dots, unlike other semiconductor laser materials, exhibit a distinct quantization energy spectrum (Tredicucci et al., 1998; Rudin et al., 2010).

Since quantum dots' peak emission frequency is very sensitive to dot structure and shape, they are beneficial. Because their activity levels are not set like the bandgap in regular bulk semiconducting, this is the case. The Exciton Bohr Radius is the overall mean separation among the electron and hole in exertions, although semiconductor crystals are frequently considerably greater than this. Whenever the diameter of the semiconductor crystal is tiny and sufficiently reaches this range, the energy states are no longer constant, and the semiconductor material is referred to as a quantum dot. The bandgap energy is changed by adding and subtracting atoms in the quantum dot crystal and modifying its surface shape. Figure 7.17 shows how quantum dots utilize this to fine-tune the emission wavelength (Saruwatari et al., 1976).

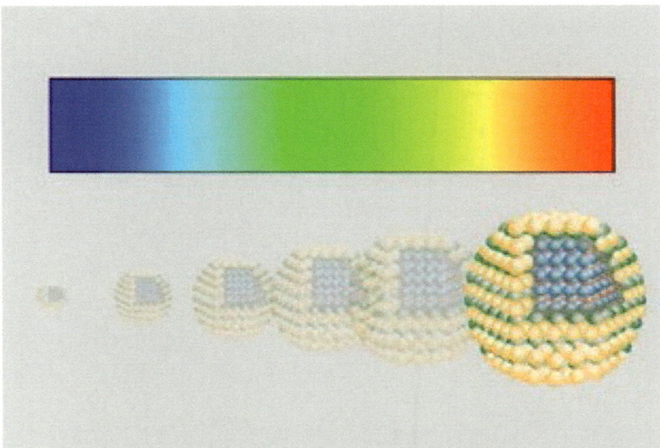

Figure 7.17. Because the bandgap widens as the dot becomes smaller, quantum dot radiation is commonly referred to as blue-shifted.

Source: https://www.researchgate.net/figure/The-band-gap-of-different-quantum-dots-and-the-relationship-between-the-band-gap-of-PbS_fig1_282293476.

Ledentsov and colleagues developed the first quantum dot laser along with a high threshold density in 1994 (Daneu et al., 1986; Hubeatir, 2020). Dots tinier than 10–20 nm as well as a dot concentration of one linked dot pair per (200 nm)^3 are required for the quantum dot laser. In quantum well lasers, the major decay process is nonradiative, releasing optical phonons rather than photons. Since only multiphonon outputs are allowed presently, radiative decay would outweigh nonradiative disintegration in quantum dot lasers, improving the laser's effectiveness. The pump will now just have to outperform the risk of loss because of the spontaneous emission, multiphoton decaying, and leaking, with the optimal threshold current being defined by the overall spontaneous emission rate. Quantum dot lasers have a lower limit power as compared to the quantum well lasers by many magnitudes (Kuang et al., 1999; Kreisler et al., 2003).

REFERENCES

1. Adachi, C., & Sandanayaka, A. S., (2020). The leap from organic light-emitting diodes to organic semiconductor laser diodes. *CCS Chemistry, 2*(4), 1203–1216.
2. Agrawal, G. P., & Dutta, N. K., (1986). *Long-Wavelength Semiconductor Lasers* (Vol. 1). New York: Van Nostrand Reinhold.
3. Alferov, Z. I., (2001). Nobel lecture: The double heterostructure concept and its applications in physics, electronics, and technology. *Reviews of Modern Physics, 73*(3), 767.
4. Anscombe, N., (2007). Join up the quantum dots. *Nature Photonics, 1*(7), 360, 361.
5. Arimitsu, T., & Shibata, F., (1983). A microscopic laser theory. I. Basic formulation. *Journal of the Physical Society of Japan, 52*(3), 772–779.
6. Balkanski, M., Wallis, R. F., & Wallis, R. F., (2000). *Semiconductor Physics and Applications* (Vol. 8, pp. 40–47). Oxford University Press.
7. Basov, N. G., Krokhin, O. N., & Popov, Y. M., (1961). Production of negative temperature states in p-n junctions of degenerate semiconductors. *Sov. Phys. JETP, 13*(6), 1320, 1321.
8. Childs, G. N., Brand, S., & Abram, R. A., (1986). Intervalence band absorption in semiconductor laser materials. *Semiconductor Science and Technology, 1*(2), 116.
9. Coldren, L. A., Fish, G. A., Akulova, Y., Barton, J. S., Johansson, L., & Coldren, C. W., (2004). Tunable semiconductor lasers: A tutorial. *Journal of Lightwave Technology, 22*(1), 193.
10. Coleman, J. J., (2012). The development of the semiconductor laser diode after the first demonstration in 1962. *Semiconductor Science and Technology, 27*(9), 090207.
11. Crowley, M. T., Naderi, N. A., Su, H., Grillot, F., & Lester, L. F., (2012). GaAs-based quantum dot lasers. In: *Semiconductors and Semimetals* (Vol. 86, pp. 371–417). Elsevier.
12. Daneu, V., DeGloria, D. P., Sanchez, A., Tong, F., & Osgood, Jr. R. M., (1986). Electron-pumped high-efficiency semiconductor laser. *Applied Physics Letters, 49*(10), 546–548.
13. De Loach, B. C., Hakki, B. W., Hartman, R. L., & D'asaro, L. A., (1973). Degradation of CW GaAs double-heterojunction lasers at 300 K. *Proceedings of the IEEE, 61*(7), 1042–1044.

14. Deacon, D. A., (1981). Basic theory of the isochronous storage ring laser. *Physics Reports, 76*(5), 349–391.
15. Dowden, J., (2009). The theory of laser materials processing. *Heat and Mass Transfer in Modern Technology*, 95–128.
16. Dumke, W., (1975). The angular beam divergence in double-heterojunction lasers with very thin active regions. *IEEE Journal of Quantum Electronics, 11*(7), 400–402.
17. Eliseev, P. G., (1996). Optical strength of semiconductor laser materials. *Progress in Quantum Electronics, 20*(1), 1–82.
18. Ettenberg, M., Nuese, C. J., & Kressel, H., (1979). The temperature dependence of threshold current for double-heterojunction lasers. *Journal of Applied Physics, 50*(4), 2949, 2950.
19. Gihleengen, B. E., Thoresen, L., & Grimsrud, K. A., (2007). Semiconductor lasers. *TFY14 Functional Materials, 2007*, 22.
20. Gmachl, C., Tredicucci, A., Sivco, D. L., Hutchinson, A. L., Capasso, F., & Cho, A. Y., (1999). Bidirectional semiconductor laser. *Science, 286*(5440), 749–752.
21. Grundmann, M., (2010). *Physics of Semiconductors* (Vol. 11, pp. 401–472). Berlin: Springer.
22. Hader, J., Moloney, J. V., & Koch, S. W., (1999). Microscopic theory of gain, absorption, and refractive index in semiconductor laser materials-influence of conduction-band nonparabolicity and coulomb-induced intersubband coupling. *IEEE Journal of Quantum Electronics, 35*(12), 1878–1886.
23. Hall, R. N., Fenner, G. E., Kingsley, J. D., Soltys, T. J., & Carlson, R. O., (1962). Coherent light emission from GaAs junctions. *Physical Review Letters, 9*(9), 366.
24. Hamaguchi, C., & Hamaguchi, C., (2010). *Basic Semiconductor Physics* (Vol. 9, pp. 443–510). Berlin: Springer-Verlag.
25. Hawkes, J., & Latimer, I., (1995). *Lasers: Theory and Practice* (Vol. 2, pp. 5–17). Prentice Hall.
26. Henry, C., (1982). Theory of the linewidth of semiconductor lasers. *IEEE Journal of Quantum Electronics, 18*(2), 259–264.
27. Herd, R. M., Dover, J. S., & Arndt, K. A., (1997). Basic laser principles. *Dermatologic Clinics, 15*(3), 355–372.

28. Hubeatir, K. A., (2020). Laser transmission welding of PMMA using IR semiconductor laser complemented by the Taguchi method and grey relational analysis. *Materials Today: Proceedings, 20*, 466–473.

29. Iga, K., Koyama, F., & Kinoshita, S., (1988). Surface-emitting semiconductor lasers. *IEEE Journal of Quantum Electronics, 24*(9), 1845–1855.

30. Kränkel, C., Marzahl, D. T., Moglia, F., Huber, G., & Metz, P. W., (2016). Out of the blue: Semiconductor laser pumped visible rare-earth doped lasers. *Laser & Photonics Reviews, 10*(4), 548–568.

31. Kreisler, M., Kohnen, W., Marinello, C., Schoof, J., Langnau, E., Jansen, B., & D'Hoedt, B., (2003). Antimicrobial efficacy of semiconductor laser irradiation on implant surfaces. *International Journal of Oral & Maxillofacial Implants, 18*(5).

32. Kroemer, H., (1963). A proposed class of hetero-junction injection lasers. *Proceedings of the IEEE, 51*(12), 1782, 1783.

33. Kuang, J. H., Sheen, M. T., Wang, S. C., Chen, C. H., & Cheng, W. H., (1999). Crack formation mechanism in laser-welded Au-coated invar materials for semiconductor laser packaging. *IEEE Transactions on Advanced Packaging, 22*(1), 94–100.

34. Lan, H., & Ding, Y., (2012). Ordering, positioning and uniformity of quantum dot arrays. *Nano Today, 7*(2), 94–123.

35. Łukasiak, L., & Jakubowski, A., (2010). History of semiconductors. *Journal of Telecommunications and Information Technology*, 3–9.

36. Manheimer, W., Colombant, D., & Goncharov, V., (2008). The development of a Krook model for nonlocal transport in laser produced plasmas. I. Basic theory. *Physics of Plasmas, 15*(8), 083103.

37. Numai, T., (2015). Fundamentals of semiconductor lasers. *Fundamentals of Semiconductor Lasers*, 89–186.

38. Ohtsubo, J., (2012). *Semiconductor Lasers: Stability, Instability and Chaos* (Vol. 111). Springer.

39. Oyama, Y., Mamada, M., Shukla, A., Moore, E. G., Lo, S. C., Namdas, E. B., & Adachi, C., (2020). Design strategy for robust organic semiconductor laser dyes. *ACS Materials Letters, 2*(2), 161–167.

40. Paschotta, R., (2020). *RP Photonics, Encyclopedia of Laser Physics and Technology* (Vol. 1, pp. 10–15). Available on: https://www.rpphotonics.com/saturation-power.html (accessed on 20 July 2022).

41. Ready, J. F., (1997). Chapter 15; Applications for surface treatment. *Industrial Applications of Lasers,* 373–383.
42. Rudin, B., Wittwer, V. J., Maas, D. J. H. C., Hoffmann, M., Sieber, O. D., Barbarin, Y., & Keller, U., (2010). High-power MIXSEL: An integrated ultrafast semiconductor laser with 6.4 W average power. *Optics Express, 18*(26), 27582–27588.
43. Samuel, I. D. W., & Turnbull, G. A., (2007). Organic semiconductor lasers. *Chemical Reviews, 107*(4), 1272–1295.
44. Samuel, I. D., & Turnbull, G. A., (2004). Polymer lasers: Recent advances. *Materials Today, 7*(9), 28–35.
45. Saruwatari, M., Kimura, T., & Otsuka, K., (1976). Miniaturized CW $LiNdP_4O_{12}$ laser pumped with a semiconductor laser. *Applied Physics Letters, 29*(5), 291–293.
46. Schubert, E. F., (2006). *Light Emitting Diodes*. Cambridge University.
47. Stenholm, S., (1986). The semiclassical theory of laser cooling. *Reviews of Modern Physics, 58*(3), 699.
48. Stern, F., & Woodall, J. M., (1974). Photon recycling in semiconductor lasers. *Journal of Applied Physics, 45*(9), 3904–3906.
49. Tredicucci, A., Gmachl, C., Capasso, F., Sivco, D. L., Hutchinson, A. L., & Cho, A. Y., (1998). A multiwavelength semiconductor laser. *Nature, 396*(6709), 350–353.
50. Tucker, R. S., (1981). Circuit model of double-heterojunction laser below threshold. *IEE Proceedings I-Solid-State and Electron Devices, 128*(3), 101–106.
51. Tucker, R. S., (1985). High-speed modulation of semiconductor lasers. *IEEE Transactions on Electron Devices, 32*(12), 2572–2584.
52. Van, R. L. J., (1984). Double heterojunction lasers and quantum well lasers. *Journal of Luminescence, 29*(5, 6), 123–161.
53. Wang, C., (2000). Laser vector measurement technique for the determination and compensation of volumetric positioning errors. Part I: Basic theory. *Review of Scientific Instruments, 71*(10), 3933–3937.
54. Wieman, C. E., & Hollberg, L., (1991). Using diode lasers for atomic physics. *Review of Scientific Instruments, 62*(1), 1–20.
55. Yablonovitch, E., & Kane, E. O., (1988). Band structure engineering of semiconductor lasers for optical communications. *Journal of Lightwave Technology, 6*(8), 1292–1299.

56. Yamada, M., (2014). *Theory of Semiconductor Lasers, 185*, 50–55.

57. Yang, P., (2005). The chemistry and physics of semiconductor nanowires. *MRS Bulletin, 30*(2), 85–91.

58. Zhao, L., Inoue, M., Yoshida, K., Sandanayaka, A. S., Kim, J. H., Ribierre, J. C., & Adachi, C., (2015). Singlet-triplet exciton annihilation nearly suppressed in organic semiconductor laser materials using oxygen as a triplet quencher. *IEEE Journal of Selected Topics in Quantum Electronics, 22*(1), 26–34.

CHAPTER 8

PHYSICS OF FERROELECTRICS

CONTENTS

8.1. INTRODUCTION

In a zero applied electric field, a ferroelectric is an insulating system having two or more distinct metastable or stable states of distinct nonzero electric polarization, known as "spontaneous" polarization. To be termed ferroelectric, a system should be able to flip between such states using an applied electric field that alters the relative energy of the states by linking the field to the polarization E • P (Webb, 2003; Rabe et al., 2007).

Electric polarization is therefore crucial to comprehending ferroelectricity. Electric polarization is simply expressed as the dipole moment, acquired from the density of charge, divided by the system volume for a finite system (Though it must be emphasized that the system boundaries are not sharp for nanosized systems). The description of polarization as a bulkiness attribute for an infinite crystal has long been more difficult to put into words (Auciello et al., 1998; Muševic et al., 2000). The answer proposed by Vanderbilt, King-Smith, and Resta, commonly known as the "contemporary polarization theory," is to recognize polarization as an integrated current that is transformed from one version to another, with the outcome stated as a Berry phase. This concentration on induced current contains the added benefit of providing a direct link to the optical and electrical polarization measurement methods. Another interesting feature of this approach is that polarization is a crystalline structure $P_0 + 2eR/\Omega$, where R is a crystal Bravais lattice vector and Ω is the unit cell volume. A formulation of the Berry-phase expression in the context of the centers of crystalline Wannier functions provides an intelligible structure-based representation (Scott, 1999; Chandra and Littlewood, 2007).

8.2. HYSTERESIS LOOPS AND SWITCHING

As previously stated, ferroelectricity is defined by the ability to transition between distinct metastable states by applying and removing an electric field. As described in more depth in the contributions by Dawber, Triscone, Lichtensteiger, and Giamarchi, Triscone, and Paruch, the mechanism of shifting is comprehended to occur on scales larger than the unit-cell scale, and often needs the development and shrinking of domains by the motion of domain walls. Whereas the "minimal distortion" crystallographic requirement outlined in the following section must allow shifting in most cases, a loop of electric hysteresis is required to produce ferroelectricity (Hardtl, 1976; Scott, 1996).

The loop of ferroelectric P–E hysteresis is symmetric in its canonical form, and the coercive fields and remainder polarization are easy to characterize and extract. Furthermore, there have been several possible problems in doing this assessment correctly (Figure 8.1) (Watanabe, 2005; Nelson et al., 2020).

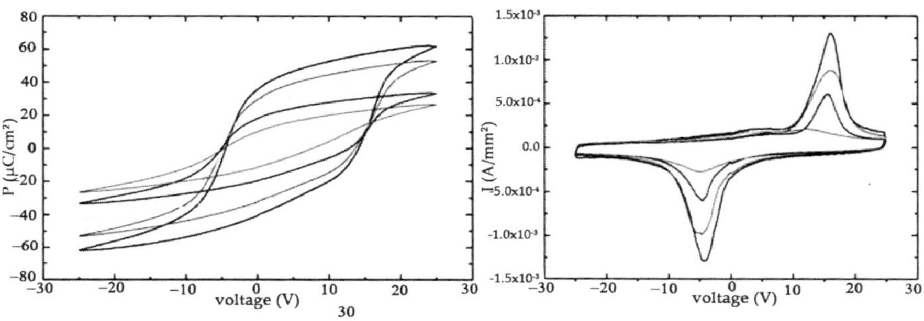

Figure 8.1. (a) Hysteresis loops for various polarizations of superlattice Pb-TiO_3/$SrTiO_3$ specimens; (b) unlike leakage current I–V curves, current-voltage loops are acquired during voltage sweeps in which the voltage is cycled.

Source: https://www.researchgate.net/figure/a-Hysteresis-loops-for-a-number-of-PbTiO3-SrTiO3-superlattice-samples-with-different_fig12_226834160.

The electrical characteristics of a ferroelectric layered may only be evaluated when it is manufactured as a device, which is the initial point for every measurement of such kind (like a capacitor). As a result, every measurement includes a comprehensive system, including electrodes, cables, and interfaces, and the electrical response is often dominated by other system elements instead of the intrinsic qualities of the material of concern. In this part, we will go over the main methods for measuring ferroelectric hysteresis, with a focus on the possibility of errors and how to overcome them (Dawber et al., 2005; Spaldin, 2007).

8.3. CRYSTALLOGRAPHIC SIGNATURE OF FERROELECTRICITY

The spontaneous polarization throughout all recognized ferroelectric crystals is caused by the array of atoms of ions in the crystalline structure, which is dependent on their locations in traditional ferroelectrics or charge sequencing of several valences in electronic ferroelectrics. Only a crystal that contains a polar space group may have a nonzero spontaneous polarization

(Hill, 2000; Vendik, 2009). Furthermore, to be considered ferroelectric, it should be feasible to flip among distinct variations with an applied electric field, which means that several polar crystals, such as wurtzite-structure insulators, are not ferroelectric. A comparatively slight symmetry-breaking deformation of a high-symmetry reference state can be used to construct the crystalline structure, which makes sure the existence of distinct states of unique polarization and improves the capacity to shift amongst them with an available electric field (Clark and Lagerwall, 1981; Resta and Vanderbilt, 2007). This involves a polar displacement of the unit cell's atoms, which can be paired with nonpolar displacement designs of the atom and the corresponding strain; in certain ferroelectric oxides, such pairing can be quite strong, resulting in piezoelectric behavior and rich temperature–pressure and corresponding strain phase illustrations, as explained below for perovskite oxides. When born effective charges Z∗e and displacements u from the reference structure are used, the magnitude of spontaneous polarization may

be approximated by the linearized approximation $\frac{1}{\Omega}\Sigma Z{*}eu$, in which Ω is the unit cell's volume. This linearized estimation in several situations provides a value that is near to the Wannier-function result and full Berry phase. In the following section, we will discuss ferroelectric oxides having low-symmetry polar and high-symmetry nonpolar structures, as well as details about the polarization of the crystalline structure, which would be provided both experimentally and computationally (Rabe and Ghosez, 2007).

With the temperature rise, many ferroelectrics move from a ferroelectric phase with many symmetry-related variants to a nonpolar paraelectric state having a single variant. The higher-symmetry reference structure in these circumstances is like the crystallographic structure seen in the paraelectric phase (Amanuma et al., 1995; Lichtensteiger et al., 2007). Ferroelectric transition temperatures have been recorded ranging from lower (1 Kelvin temperature) to higher (over 1,000 Kelvin temperature); in the latter instances, the material may melt earlier than the transition temperature has been recorded. The symmetry-breaking relationship between the ferroelectric structure and paraelectric structure with high-symmetry is comparable with a 2nd-order transition and it may be characterized using Landau theory with polarization as the fundamental order parameter. This analysis concludes that the dielectric susceptibility deviates at the changeover. The Lyddane–Sachs–Teller relationship is then used to connect to the disappearing frequency of a polar phonon, which has been the core principle of soft-mode ferroelectrics theory (Okuyama and Ishibashi, 2005; Yamauchi et al., 2017).

8.4. MATERIALS

Whereas ferroelectricity had first been observed in materials that are hydrogen-bonded like Rochelle salt and, later, KDP, the findings of ferroelectricity in the considerably simpler, nonhydrogen-containing perovskite oxide $BaTiO_3$ in 1949 revolutionized our understanding of the phenomena. $BaTiO_3$ is the most well-known and studied member of the family of perovskite oxide, which comprises not only perovskite compounds but also organized and disorganized solid solutions. A greater understanding of the origins of ferroelectricity, and also first-principles modeling and quantitative phenomenology, were made possible as a result of the relative simplicity of the perovskite structure, which will be discussed in the following part (Miller and McWhorter, 1992; Dmitriev et al., 2001). Perovskites have been closely connected to other families of significant ferroelectric materials, which are intergrowths of perovskite films having other oxides. Ultimately, we would discuss other structural families of ferroelectric oxides which are not linked to perovskites (Ishibashi and Orihara, 1995; Yan et al., 2011).

8.4.1. Perovskite Oxides

The perovskite oxides are the ferroelectric oxides that have received the most attention in recent years (from the mineral perovskite $CaTiO_3$). There are several compositions in the ABO_3 family, each one represents a cation element or a combination of two or many such vacancies or elements with A and B each representing one of these elements or vacancies. There are several types of magnetic and structural order in the entire family, which are dependent on the cationic ordering and composition (Landauer et al., 1956; Naidu, 1990). They may be insulating or metallic and show a wide range of magnetic and structural order. The oxides of perovskite that have been ferroelectrics in massive crystalline structures belong to a subfamily; other associated compounds may tend to ferroelectric instability that is not manifested in the majority crystal due to the influence of other, non-ferroelectric.

Pm_3m is the space group of several ferroelectric perovskite oxides, which have a simple cubic crystalline structure and a five-atom basis. The ideal perovskite structure is the reference structure having high symmetry, as well as the structure of several ferroelectric perovskite oxides' raised temperature paraelectric phase. If the B atom is in the corner, the A atom will be in the center, and the O atoms will be at the halfway point of every edge. This means that the A atom will be in one of the corners of the cube, the B atom will be in one of the centers of each face, and oxygen will be at the halfway point of each face if the B atom is in the corner (Figures 8.2 and 8.3) (Jung et al., 1990; Bennett and Torre, 2005).

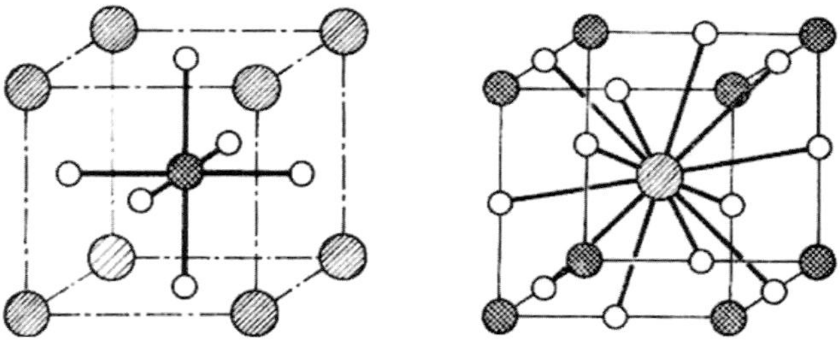

Figure 8.2. Two distinct perspectives of the ABO_3 perfect cubic perovskite structure's unit cell. An octahedron made up of oxygen atoms (white pattern) contains the B atom (grilled design). The A atom (dashed pattern) has 12 oxygen, initial neighbors.

Source: https://www.researchgate.net/figure/Two-different-views-of-the-unit-cell-of-the-ABO-3-ideal-cubic-perovskite_fig3_226834160.

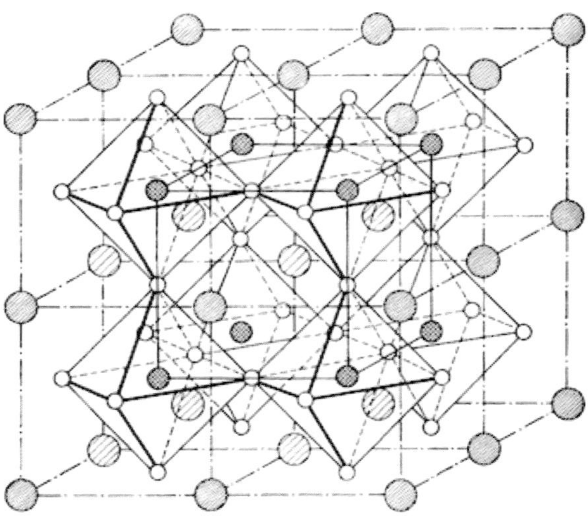

Figure 8.3. Another perspective of the perfect cubic ABO_3 perovskite structure.

Source: https://www.researchgate.net/figure/Another-view-of-the-ABO-3-ideal-cubic-perovskite-structure-From-20_fig4_226834160.

8.4.2. Other Families of Ferroelectric Oxides

The presence of various other ferroelectric oxide families has long been known. Boracites, tungsten bronzes, $Pb_5Ge_3O_{11}$, Mn_3TeO_6, $Gd_2(MoO_4)_3$, $LiNaGe_4O_9$, $R_3Sb_5O_{12}$, $Li_2Ge_7O_{15}$, and $BaAl_2O_4$ are among the types (Figure 8.4) (Bouregba and Poullain, 2002; Kumar et al., 2011).

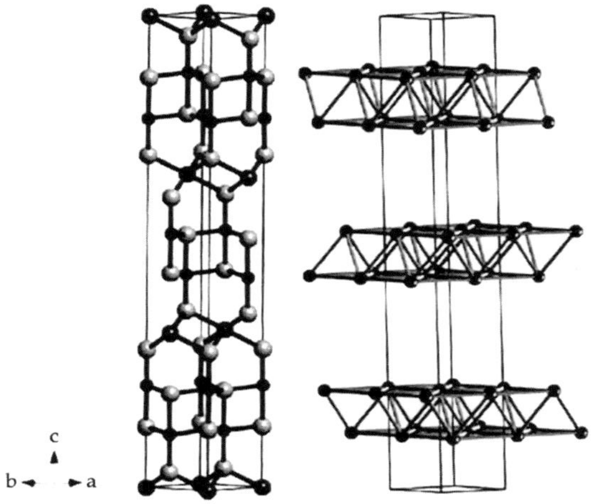

Figure 8.4. The layered array of Lu (large dark-grey spheres), Fe (small black spheres), and oxygen (big white spheres) along the c-axis in the crystal structure of $LuFe_2O_4$ (left). The Fe double layers are connected in a triangle pattern (right).

Source: https://www.researchgate.net/figure/Color-online-a-Monoclinic-crystal-structure-C2-m-of-LuFe2O4-a-595A95A-b_fig1_225288842.

With the fast development of conceptual approaches and experimental methods for studying and comprehending the physics of ferroelectric oxides, a reexamination of such families can reveal materials with new mechanisms for ferroelectricity and also before unrecognized members of such families with attractive material characteristics. KNO_3, whose crystalline structure is dependent on the triangular stacking of NO_3 units containing potassium ions, is recently the topic of both conceptual (Beteille and Livage, 1998) and experimental research, expanding on prior findings such as Chua (2019) and Prins et al. (1996). Such complexity of structure, as a result of the ferroelectric deformation pairing to the rotation of the triangular units, has a strong bond with the dense pressure-temperature phase diagram.

8.4.3. Oxides of Magnetic Ferroelectric

The hexagonal manganites of ferroelectric, illustrated by $YMnO_3$, have caught the interest of researchers owing to their magnetic characteristics, and the characteristics and structure of these materials are described in detail in Spaldin's contribution to this volume. The high-temperature paraelectric phase, in contrast to the perovskite structure, is unique in the material's evolution. It appears to be an unsuitable ferroelectric transition in this situation, the rotation of the oxygen polyhedra is assumed to be the key ordering parameter for the transition., which breaks the symmetry and results in the growth of a nonzero polarization along with c. The creation of magnetism is due to the Mn cations, which have been quite passive in the transition of ferroelectric (Holzer et al., 2005; Guo et al., 2020).

8.4.4. Electronic Ferroelectrics

A new focus has lately been drawn on the prospect of electronic ferroelectrics, which, if proven to be true, will display an electronic ground state which is symmetrically unstable in principle, even though the ions are held stationary. Several theoretical studies have been conducted using the Falicov–Kimball model (Indlekofer and Kohlstedt, 2005; Zhang et al., 2007), which applies to rare-earth compounds and mixed-valent transition-metal, like SmB_6, and is shown to be effective. Charge-ordering events (such as charge-density waves (CDWs)) in complicated oxides, such as manganites with a perovskite structure, are another mechanism for electronic ferroelectricity that has been proposed. In addition to transition metals, a few additional elements may exhibit charge disproportionation, a type of charge order that is more common than multiple valences in transition metals. The presence of 2, Bi_4^{+}, Bi_3^{++}, and Bi_5^{+} in $BaBiO_3$, for instance, is reported to induce a cell-doubling transition, which is further examined in Rabe and Ghosez's work. The charge-ordered states in this scenario, as in the majority of previous situations, have a reasonably high degree of symmetry despite the charge ordering. Several systems possessing spontaneous polarization (such as $LuFe_2O_4$ and YFe_2O_4) (Shrout and Fielding, 1990; Mikolajick et al., 2021) are reported, which were followed by a significant dielectric response above the phase-transition temperature. Specifically, the magnetic and ferroelectric characteristics of $LuFe_2O_4$, as depicted in Figure 8.4, have recently generated a great deal of research. It is also observed that Fe_2BO_4 exhibits an intriguing dielectric characteristic connected to charge sequencing. The large dielectric invariant of $La_{1.5}Sr_{0.5}NiO_4$ is connected to CDWs; the association between CDW systems and electronic ferroelectric has been elaborated on in Littlewood and Chandra's work (Shen et al., 2019; Zhu et al., 2021).

8.4.5. Micro-Ferroelectrics

The way we think about ferroelectricity in systems including one or more decreased parameters has changed significantly over the last decade. This was generally believed till the late 1990s that ferroelectricity in perovskite oxides will vanish below a critical size of roughly 10 nm. Since then, a new perspective has appeared based on a combination of theoretical and experimental research that this restriction is an external effect caused by mechanical and electrical boundary circumstances instead of intrinsic size effects linked to the ferroelectric instability's collective nature.

The experimental discovery is that PZT layers may sustain switchable polarization ordinary to the layer down to thicknesses of several nanometers. As the depolarization field was suitably balanced, this observation was corroborated by theoretical assumptions that $PbTiO_3$ would sustain switchable spontaneous polarization regular to the layer down to thicknesses of fewer than 3-unit cells. This prompted a focus on the depolarization field's role in generating perceptible size effects, with Dawber et al. (2005) showing that the evaluated coercive field's anticipated thickness reliance is due to constant voltage drops in realistic electrodes, and multiple studies have on the impact of imperfect field remuneration. Inadequate compensation can also lead to the creation of complicated domain structures, in addition to reducing ferroelectricity. New questions arise as a result of the potential of ferroelectricity in nanosized systems. Technically, there has been no ferroelectric paraelectric phase transition in a confined system, hence the condition for a sustained switchable polarization will have to be expressed dynamically. Another question is whether finite systems, like layers or particles, can exist in asymmetric surroundings (such as various bottom and top electrodes for layers). Even though there are numerous metastable states, their degrees of stability may vary. The stability of two or more variations having distinct polarizations and the capacity to switch from one to the other through an applied electric field ought to be the condition for ferroelectricity in both circumstances (Das and Appenzeller, 2011).

Current basic and applied study in the physics of non-bulk ferroelectrics has yielded several significant findings, only a few of which are highlighted here. The following part (on applications), the contribution by Ghosez and Rabe (on 1st-principles theoretical investigations), and the contribution by

Triscone, Dawber, and Lichtensteiger, provide more discussion (on finite-size effects).

8.5. USAGES OF FERROELECTRIC MATERIALS

Ferroelectric materials have sparked a lot of attention as they have features that are useful for a variety of purposes. In certain situations, ferroelectrics are amongst the most popular materials for a given application; the possibilities offered by innovative geometries and materials are just beginning to be explored. Here is a quick summary of both existing and potential applications.

8.5.1. Piezoelectric and Pyroelectric Devices

Several applications for the materials of ferroelectric rely on the related features of piezoelectricity and pyroelectricity rather than the ferroelectricity itself. Such qualities are needed for ferroelectricity to occur due to symmetry (They are both reasonable and necessary, in that all ferroelectrics are piezoelectric and pyroelectric, although not all piezoelectrics and pyroelectrics are ferroelectric). Pyroelectrics have a wide spectrum of uses in identification and imaging. Piezoelectrics are used in a variety of electronic devices. The strong piezoelectric coefficients of ferroelectric solid solutions like PZT enable innovative small electromechanical devices, especially in MEMS. Muralt's study provides more information on such applications (Zhang et al., 2009).

8.5.2. Ferroelectric Memory Technology

As evidenced by the present popularity of FLASH-based technology in MP3 players, mobile phones, digital cameras, and a variety of other portable gadgets, the use of nonvolatile memory technology in computation usages, particularly in mobile devices, is long apparent. Due to limits in durability, scalability, and voltage needs, FLASH has not preferred durable nonvolatile memory.

Ferroelectrics have been regarded as a leading possibility for the next development of nonvolatile memories for about two decades. Commonly accessible gadgets exist as independent higher-capacity nonvolatile memory; nevertheless, they appear to be more prevalent in special applications or as part of an integrated memory gadget that combines technologies having multiple memory than nonvolatile memory devices having a standalone higher capacity. MRAMs and phase-change memories are two further technologies that are equivalent and have already been commercialized.

It has been difficult to use ferroelectrics as memory elements because one of the issues has been that to find out the ferroelectric's polarization state, it is essential to switch the polarization; this causes the stored data to be erased and then rewritten, implying that the read-out operation has been catastrophic. Because ferroelectrics may become fatigued after repeated cycling, this catastrophic read-out harms the reliability of the system, but fatigue resistance has been increased over time through the usage of oxide electrodes. As well of interest are noncatastrophic read-out gadgets, such as ferroelectric field-effect transistors, as well as the usage of ferroelectric and ferromagnetic alignments paired materials or composite structures, wherein the direction of electrical polarization may be determined by measuring the magnetic field. The contributions by Spalding in this edition as well as a current study provide extra details on multiferroics (Viswanath et al., 2020).

8.5.3. Ferroelectric Nanostructures

Ferroelectrics have been produced in a variety of innovative shapes in recent times, which has led to the development of novel applications.

Ferroelectric materials tend to keep their characteristics amazingly well when their size is reduced, with ferroelectric nanoparticles retaining ferroelectric until at a minimum of 20 nanometers. In ferroelectric thin layers, the critical thickness is on the order of several unit cells. (see Dawber, Triscone, and Lichtensteiger's contribution). As a result, high-density arrangements of ferroelectric nano capacitors might be used to create exceptionally higher-capacity memory systems. Capacitors of nanoscale dimensions may be made via e-beam direct writing and focused-ion-beam (FIB) milling. Moreover, a major challenge remains in producing such high-density arrays rapidly and cost-effectively while preserving acceptable registries and characteristics of the material, which has sparked interest in ferroelectric nanocapacitor self-patterning.

Several companies have developed ferroelectric nano-shell tubes, which are polycrystalline ceramic tubes having nanoscale walls. These tubes have a considerable promise for a variety of applications, including high dimensional memory and microfluidic delivery systems. The tubes are produced by growing them in a porous matrix, which improves their homogeneity and registry. The FIB method, which is used by the Belfast group to cut solitary crystals into thin layers, rings, and other fascinating structures, is another method for producing ferroelectrics with unique nanoscale structures that are producing promising outcomes (Krowne et al., 2011).

8.5.4. Field-Effect Devices

The notion of employing ferroelectric field-effect transistors, wherein the gate dielectric of a standard FET is substituted with the material of ferroelectric, was briefly described before. The ferroelectric's screenings charges change the semiconductor channel's conductivity. The primary barrier to overcome is the low retention times characteristic of such devices; this is an important topic of study, described in more depth elsewhere. The ferroelectric field effect may also be utilized to alter the characteristics of more unusual systems, such as superconductors and magnetic oxides (Nguyen and Scheick, 2001).

REFERENCES

1. Amanuma, K., Hase, T., & Miyasaka, Y., (1995). Preparation and ferroelectric properties of $SrBi_2Ta_2O_9$ thin films. *Applied Physics Letters, 66*(2), 221–223.
2. Auciello, O., Scott, J. F., & Ramesh, R., (1998). The physics of ferroelectric memories. *Physics Today, 51*(7), 22–27.
3. Bennett, L. H., & Della, T. E., (2005). Analysis of wasp-waist hysteresis loops. *Journal of Applied Physics, 97*(10), 10E502.
4. Beteille, F., & Livage, J., (1998). Optical switching in VO_2 thin films. *Journal of Sol-Gel Science and Technology, 13*(1), 915–921.
5. Bouregba, R., & Poullain, G., (2002). Numerical extraction of the true ferroelectric polarization due to switching domains from hysteresis loops measured using a sawyer-tower circuit. *Ferroelectrics, 274*(1), 165–181.
6. Chandra, P., & Littlewood, P. B., (2007). A landau primer for ferroelectrics. In: *Physics of Ferroelectrics* (pp. 69–116). Springer, Berlin, Heidelberg.
7. Chua, L., (2019). Resistance switching memories are memristors. In: *Handbook of Memristor Networks* (pp. 197–230). Springer, Cham.
8. Clark, N. A., & Lagerwall, S. T., (1981). Physics of ferroelectric fluids: The discovery of a high-speed electro-optic switching process in liquid crystals. In: *Recent Developments in Condensed Matter Physics* (pp. 309–319). Springer, Boston, MA.
9. Das, S., & Appenzeller, J., (2011). FETRAM. An organic ferroelectric material based novel random access memory cell. *Nano Letters, 11*(9), 4003–4007.
10. Dawber, M., Rabe, K. M., & Scott, J. F., (2005). Physics of thin-film ferroelectric oxides. *Reviews of Modern Physics, 77*(4), 1083.
11. Dmitriev, S. V., Semagin, D. A., Abe, K., Vasiliev, A. A., & Shigenari, T., (2001). Elastically hinged molecule model in physics of ferroelectric materials. *Ferroelectrics, 262*(1), 53–58.
12. Guo, W., Liu, X., Han, S., Liu, Y., Xu, Z., Hong, M., & Sun, Z., (2020). Room-temperature ferroelectric material composed of a two-dimensional metal halide double perovskite for x-ray detection. *Angewandte Chemie International Edition, 59*(33), 13879–13884.

13. Hardtl, K. H., (1976). Physics of ferroelectric ceramics used in electronic devices. *Ferroelectrics, 12*(1), 9–19.
14. Hill, N. A., (2000). Why are there so few magnetic ferroelectrics?. *The Journal of Physical Chemistry B, 104*(29), 6694–6709.
15. Holzer, F., Kopinke, F. D., & Roland, U., (2005). Influence of ferroelectric materials and catalysts on the performance of non-thermal plasma (NTP) for the removal of air pollutants. *Plasma Chemistry and Plasma Processing, 25*(6), 595–611.
16. Indlekofer, K. M., & Kohlstedt, H., (2005). Simulation of quantum dead-layers in nanoscale ferroelectric tunnel junctions. *EPL (Europhysics Letters), 72*(2), 282.
17. Ishibashi, Y., & Orihara, H., (1995). A theory of DE hysteresis loop. *Integrated Ferroelectrics, 9*(1–3), 57–61.
18. Jung, P., Gray, G., Roy, R., & Mandel, P., (1990). Scaling law for dynamical hysteresis. *Physical Review Letters, 65*(15), 1873.
19. Krowne, C. M., Kirchoefer, S. W., Chang, W., Pond, J. M., & Alldredge, L. M. B., (2011). Examination of the possibility of negative capacitance using ferroelectric materials in solid state electronic devices. *Nano Letters, 11*(3), 988–992.
20. Kumar, A., Ovchinnikov, O. S., Funakubo, H., Jesse, S., & Kalinin, S. V., (2011). Real-space mapping of dynamic phenomena during hysteresis loop measurements: Dynamic switching spectroscopy piezoresponse force microscopy. *Applied Physics Letters, 98*(20), 202903.
21. Landauer, R., Young, D. R., & Drougard, M. E., (1956). Polarization reversal in the barium titanate hysteresis loop. *Journal of Applied Physics, 27*(7), 752–758.
22. Lichtensteiger, C., Dawber, M., & Triscone, J. M., (2007). Ferroelectric size effects. In: *Physics of Ferroelectrics* (pp. 305–338). Springer, Berlin, Heidelberg.
23. Mikolajick, T., Slesazeck, S., Mulaosmanovic, H., Park, M. H., Fichtner, S., Lomenzo, P. D., & Schroeder, U., (2021). Next generation ferroelectric materials for semiconductor process integration and their applications. *Journal of Applied Physics, 129*(10), 100901.
24. Miller, S. L., & McWhorter, P. J., (1992). Physics of the ferroelectric nonvolatile memory field effect transistor. *Journal of Applied Physics, 72*(12), 5999–6010.

25. Muševic, I., Blinc, R., & Zekš, B., (2000). The physics of ferroelectric and antiferroelectric liquid crystals. *World Sci., 46*, 215–222.
26. Naidu, S. R., (1990). Simulation of the hysteresis phenomenon using Preisach's theory. *IEE Proceedings A-Physical Science, Measurement and Instrumentation, Management and Education, 137*(2), 73–79.
27. Nelson, C. T., Vasudevan, R. K., Zhang, X., Ziatdinov, M., Eliseev, E. A., Takeuchi, I., & Kalinin, S. V., (2020). Exploring physics of ferroelectric domain walls via Bayesian analysis of atomically resolved STEM data. *Nature Communications, 11*(1), 1–12.
28. Nguyen, D. N., & Scheick, L. Z., (2001). TID testing of ferroelectric nonvolatile RAM. In: *2001 IEEE Radiation Effects Data Workshop. NSREC 2001. Workshop Record. Held in conjunction with IEEE Nuclear and Space Radiation Effects Conference (Cat. No. 01TH8588)* (pp. 57–61). IEEE.
29. Okuyama, M., & Ishibashi, Y., (2005). *Ferroelectric Thin Films: Basic Properties and Device Physics for Memory Applications* (Vol. 98). Springer Science & Business Media.
30. Prins, M. W. J., Grosse-Holz, K. O., Müller, G., Cillessen, J. F. M., Giesbers, J. B., Weening, R. P., & Wolf, R. M., (1996). A ferroelectric transparent thin-film transistor. *Applied Physics Letters, 68*(25), 3650–3652.
31. Rabe, K. M., & Ghosez, P., (2007). First-principles studies of ferroelectric oxides. In: *Physics of Ferroelectrics* (pp. 117–174). Springer, Berlin, Heidelberg.
32. Rabe, K. M., Dawber, M., Lichtensteiger, C., Ahn, C. H., & Triscone, J. M., (2007). Modern physics of ferroelectrics: Essential background. In: *Physics of Ferroelectrics* (pp. 1–30). Springer, Berlin, Heidelberg.
33. Resta, R., & Vanderbilt, D., (2007). Theory of polarization: A modern approach. In: *Physics of Ferroelectrics* (pp. 31–68). Springer, Berlin, Heidelberg.
34. Scott, J. F., (1996). Device physics of ferroelectric memories. *Ferroelectrics, 183*(1), 51–63.
35. Scott, J. F., (1999). Device physics of ferroelectric thin-film memories. *Japanese Journal of Applied Physics, 38*(4S), 2272.
36. Shen, H., Liu, J., Chang, K., & Fu, L., (2019). In-plane ferroelectric tunnel junction. *Physical Review Applied, 11*(2), 024048.

37. Shrout, T. R., & Fielding, J., (1990). Relaxor ferroelectric materials. In: *IEEE Symposium on Ultrasonics* (pp. 711–720). IEEE.
38. Spaldin, N. A., (2007). Analogies and differences between ferroelectrics and ferromagnets. In: *Physics of Ferroelectrics* (pp. 175–218). Springer, Berlin, Heidelberg.
39. Vendik, O. G., (2009). Ferroelectrics find their "niche" among microwave control devices. *Physics of the Solid State, 51*(7), 1529–1534.
40. Viswanath, P., De Silva, K. K. H., Huang, H. H., & Yoshimura, M., (2020). Large piezoresponse in ultrathin organic ferroelectric nano lamellae through self-assembly processing. *Applied Surface Science, 532*, 147188.
41. Watanabe, Y., (2005). Physics of ferroelectric interfaces: An attempt at nanoferroelectric physics. *Ferroelectric Thin Films,* 177–199.
42. Webb, J. F., (2003). On the physics of ferroelectrics. *Science Progress, 86*(3), 203–234.
43. Yamauchi, K., Barone, P., & Picozzi, S., (2017). Topological phase transition coupled with spin-valley physics in ferroelectric oxide heterostructures. *Physical Review B, 95*(3), 035146.
44. Yan, H., Inam, F., Viola, G., Ning, H., Zhang, H., Jiang, Q., & Reece, M. J., (2011). The contribution of electrical conductivity, dielectric permittivity and domain switching in ferroelectric hysteresis loops. *Journal of Advanced Dielectrics, 1*(01), 107–118.
45. Zhang, S., Li, F., Sherlock, N. P., Luo, J., Lee, H. J., Xia, R., & Shrout, T. R., (2011). Recent developments on high curie temperature PIN–PMN–PT ferroelectric crystals. *Journal of Crystal Growth, 318*(1), 846–850.
46. Zhang, S., Lim, J. B., Lee, H. J., & Shrout, T. R., (2009). Characterization of hard piezoelectric lead-free ceramics. *IEEE Transactions on Ultrasonics, Ferroelectrics, and Frequency Control, 56*(8), 1523–1527.
47. Zhang, S., Xia, R., & Shrout, T. R., (2007). Modified $(K_{0.5}Na_{0.5})$ Nb O_3 based lead-free piezoelectrics with broad temperature usage range. *Applied Physics Letters, 91*(13), 132913.
48. Zhu, X., Fan, W., Ren, W., & Li, Y., (2021). Mechanisms of molecular ferroelectrics made simple. *The Journal of Physical Chemistry C, 125*(22), 12461–12467.

INDEX

F

G

H

I

K

L

M

T

V

W